OXFORD READINGS IN PH

THE PHILOSOPHY OF TIME

THE PHILOSOPHY
OF TIME

Edited by

ROBIN LE POIDEVIN
and
MURRAY MACBEATH

OXFORD UNIVERSITY PRESS

BD
638
.P49
1993

OXFORD
UNIVERSITY PRESS

Great Clarendon Street, Oxford OX2 6DP

Oxford University Press is a department of the University of Oxford.
It furthers the University's objective of excellence in research, scholarship,
and education by publishing worldwide in

Oxford New York

Auckland Bangkok Beunos Aieres Cape Town Chennai
Dar es Salaam Delhi Hong Kong Istanbul Karachi Kolkata
Kuala Lumpur Madrid Melbourne Mexico City Mumbai Nairobi
São Paulo Shanghai Taipei Tokyo Toronto

Oxford is a registered trade mark of Oxford University Press
in the UK and in certain other countries

Published in the United States
by Oxford University Press Inc., New York

Introduction and selection © Oxford University Press 1993

First published in paperback 1993

British Library Cataloguing in Publication Data

Data available

Library of Congress Cataloging in Publication Data

The Philosophy of time / edited by Robin Le Poldevin and Murray MacBeath.
p. cm.—Oxford readings in philosophy)
Includes bibliographical references and index.
1. Time. I. Le Poldevin, Robin, 1962– . II. Macbeath, Murray. III. Series.
BD638.P49 1993 115—dc20 92–26125

ISBN 0–19–823999–8 (Pbk.)

13

Printed in Great Britain
on acid-free paper by
Biddles Ltd., King's Lynn, Norfolk

CONTENTS

INTRODUCTION

ROBIN LE POIDEVIN and MURRAY MACBEATH

Consider three fundamental beliefs we have about the world (so fundamental that we would rarely, if ever, articulate them): that change is going on constantly, that changes are caused, and that there are constraints on what changes are possible. If we then ask: but are these beliefs *true*? and: how is it *possible* for them to be true, if they are? we have summarized many of the central concerns of metaphysics, the philosophical study of what there is. These are the questions with which the first philosophers were concerned. Such questions are paradigmatic of philosophical enquiry, and one cannot progress very far in answering them without considering problems about time. One of the purposes of this Introduction is to make those connections clear.

The three fundamental beliefs mentioned above introduced three central concerns: change, causation, and possibility. These concerns are unifying themes in the essays of this volume.

CHANGE AND THE PASSAGE OF TIME

It is a commonplace that time, not space, is the dimension of change. There is a wholly uncontroversial sense in which this is true: genuine change involves temporal variation in the ordinary properties of things: a hot liquid cools, a tree blossoms, an iron gate rusts. Purely *spatial* variation, for example the distribution of various colours in a patterned rug, does not count as genuine change. Uncontroversial as this is, it requires explanation. What is special about time? One persuasive answer to this is provided by those who think that 'time is the dimension of change' is true in another sense: that times themselves change in that what is future will become present and then recede further and further into the past. This, the so-called 'passage' or 'flow' of time, is a much more controversial issue, and is a central topic in Part 1. Believers in time's passage can say that time, not space, is the dimension of ordinary change (i.e. change in the properties of things) because time passes in a way in which space does not. Temporal passage makes ordinary change possible, for this reason: if the passage of time were not to bring with it from the future into the present the event which constitutes the tree's blossoming, the tree could not change from being in bud to being in blossom.

Talk of time's passage is to some extent metaphorical, as the notion it invokes, movement, is applied to things in time. Serious talk of a *time* moving is in danger of committing a category mistake. What, then, is the philosophical content of the metaphor of passage? Arthur Prior addresses this question in the second essay of this volume, and shows in what ways the metaphor can mislead.

The issue over whether time passes or not can be represented as an issue over what makes tensed statements true. By 'tensed statements' we mean those that reflect a given temporal perspective: this reflection may manifest itself in the grammatical tense of a verb or in the use of some such term as 'past', 'present', 'future', 'now', 'then', 'yesterday', 'next year', 'the month after next', 'in a short while', 'a long time ago', etc. Suppose you say at four o'clock: 'It is now four o'clock'. What makes this statement true? There are two accounts we can give. On the first account (call this the 'A account'), what makes your utterance true is that a particular time, given the name 'four o'clock', is *present*, and that fact is a fact about time, not about the location of your utterance. It is also a fleeting fact, for soon four o'clock will no longer be present. On the second account (call this the 'B account'), your statement is made true by the *relational* fact that the utterance is made at four o'clock. There is no *further* fact that four o'clock is present, and the fact that the utterance is made at four o'clock is not a fleeting fact: it remains the case at all times that that particular utterance of yours is made at four o'clock. Although, on the B account, times do not become past, it can account for the truth of your next utterance, 'four o'clock will soon be past', in the following way: there is a group of times, later than the time of the utterance, such that four o'clock is earlier than those times. This is a complex relational fact.

Because proponents of the A account believe that tensed statements reflect a feature of the world that is not captured by any tenseless statement, the facts that they regard as making tensed statements true are often called 'tensed facts'. Proponents of the B account deny the existence of tensed facts.

SUBJECTIVE FACTS

It might be thought that, if there are tensed facts, there are analogous facts concerning space and persons. Imagine a once omniscient being who has a sudden attack of selective amnesia, but is otherwise very well informed. She has an unsurpassed knowledge of history: she knows all the various events that occur in each of the times of the whole time-series. She does *not* know, however, which date is the *present* one. Likewise, she has an unsurpassed knowledge of places and their distances from one another, but she doesn't know which place *she* is occupying. Finally, she knows the names and characteristics of every single person, but *doesn't* know which *she* is. One is tempted to say that there are three facts which she is not apprised of: which time is 'now', which place is 'here', and which person is 'I'. These we might call 'subjective facts'. Some philosophers, for example Richard Swinburne and Ernest Sosa,[1] have argued that there are indeed such facts, temporal, spatial, and personal subjective facts. 'Temporal subjective facts' would just be another name for tensed facts. But most proponents of tensed facts would reject the notion of spatial and personal subjective facts: tensed facts, they would argue, are precisely what make time *distinct* from space: time, not space, passes. They would embrace a relational analysis of spatially indexical statements—i.e. those statements involving such terms as 'here', 'there', 'five miles to the south-west', etc. For example, if you uttered 'It's snowing here', then your statement would be true if and only if it were snowing at the place the utterance was made. Such philosophers, believing in tensed facts, but rejecting their spatial (or personal) counterparts, would argue that Swinburne and Sosa simply obscure the crucial disanalogies between time and space.

Subjective, or indexical, beliefs, i.e. those expressed by indexical statements such as 'I am a secret agent', 'It's now time to get going', 'There will be an explosion here shortly', are a crucial part of our mental lives. They play an important role in the explanation of action, for example. Indeed, in this context, as Perry pointed out,[2] such beliefs seem to be ineliminable. John moves away from the building because he believes that there is a bomb in *this* building which is due to explode very soon. This belief cannot be equivalent to the non-indexical belief that there is a bomb in the public library which is due to go off at four o'clock, even though this is the public library and it is very nearly four o'clock, because John could have this non-indexical belief and not move a muscle (he might think this building is the opera house, or think that the time is only two o'clock). Now, unless he or she wants to follow Swinburne and Sosa, the proponent of tensed facts cannot argue from the ineliminability of tensed beliefs to the existence of tensed facts. For there would be then no reason to reject the argument from the ineliminability of spatially indexical beliefs to spatially subjective facts.

---

[1] Richard Swinburne, 'Tensed Facts', *American Philosophical Quarterly*, 27 (1990): 117–30; Ernest Sosa, 'Consciousness of the Self and of the Present', in James E. Tomberlin (ed.), *Agent, Language, and the Structure of the World* (Indianapolis: Hackett, 1983), 131–43.
[2] John Perry, 'The Problem of the Essential Indexical', *Noûs*, 13 (1979): 3–21.

So, for most philosophers, the motivation for believing in tensed facts does not come from the ineliminability of tensed beliefs. For this reason, and the fact that there already exists a volume in this series on demonstratives,[3] which includes a section on 'reality and the present', we have not included work on tensed belief in this volume. The reader is referred to the Bibliography for some references to such work.

THE REALITY OF THE FUTURE

The motivation for tensed facts, rather, comes from such considerations as these: (1) only such facts can apparently allow for genuine change; (2) time just seems to pass; (3) there are, we are tempted to think, ontological differences between the past and the future: past individuals are as real as presently existing ones, even though past individuals don't, obviously, exist in the present. In contrast, future individuals, i.e. those yet unborn, cannot be thought of as real. In addition, statements about the past can be thought to have a determinate truth-value; not so statements about the future.

Consideration (3) suggests another interpretation of the passage of time: as new people are born, events occur, changes go on, reality, as it were, expands—more facts are added to the totality of facts. This is what is meant by 'becoming'. It is not in opposition to what we called the A account of tensed statements. In fact, anyone who believes in temporal becoming must also believe the A account. However, the A account has no implications for temporal becoming: one could intelligibly hold that future individuals were real and that statements about the future had a determinate truth-value while still adopting the A account. There are further positions consistent with the A account: past and future may be equally unreal.

THE UNREALITY OF TIME

Turning now to the B account, what reason could anyone have for rejecting tensed facts? The reason most often given by those who espouse the B account is that the supposition that there are tensed facts leads to contradiction. McTaggart's classic argument for this conclusion is given in the first essay of this volume. McTaggart's position is unusual in that, although he rejected tensed facts as incoherent, he also thought that the reality of time required such facts. The inevitable conclusion he drew was that time is

---

[3] Palle Yourgrau (ed.), *Demonstratives* (Oxford: Oxford University Press, 1990).

unreal. Few philosophers have followed him in this, but those who have engaged in the debate have, generally speaking, *either* agreed with McTaggart that time entails tensed facts but rejected his argument against such facts, *or* accepted his argument (or some reconstruction) against tensed facts but tried to give an account of time and its features without recourse to such facts. Prior belongs to the first group of philosophers. Hugh Mellor, who contributes the third essay in this volume, belongs to the second group. The challenge faced by philosophers of this second group is how to account for such obvious features of the world as the direction of time. This is an issue to which we shall return later in this Introduction.

TIME WITHOUT CHANGE

The topic of change is also intimately bound up with the subject of Part 2, temporal relationism. Relationism regards the relationship between time and change to be a particularly intimate one. Beyond this central (if vague) doctrine, 'relationism' encompasses a number of different theories. The simplest view is that time just *is* change, a view which Aristotle ascribes to some of his predecessors before replacing it with his own view in book IV of the *Physics* that time is the *measure* of change. In Leibniz we find the view that times are constructions from events and their relations.[4] A given instant, on this view, is just a collection of simultaneous events. This of course entails the impossibility of empty times, that is, periods of time in which nothing changes. This is not generally thought to be an unpalatable consequence, since empty time has been rejected on other grounds. The standard argument against it, due to Aristotle, is that we cannot make sense of empty time because nothing could count as an experience of it,[5] since for me to notice time is precisely for me to undergo change. Now does this mean that the notion of empty time is *devoid of empirical content*, that is, that no data we could obtain would support the hypothesis that we lived in a world in which there were temporal vacua over the hypothesis that we didn't live in such a world? If so, then it could be argued that we have no good reason for admitting the possibility of empty time. However, Aristotle's uncontroversial point does not necessarily establish that empty time is an empirically contentless supposition, as Sydney Shoemaker shows in his essay. Shoemaker constructs a situation in which we would appear to be justified *on empirical*

---

[4] See *The Leibniz–Clarke Correspondence*, ed. H. G. Alexander (Manchester: Manchester University Press, 1956), 26–7.
[5] Aristotle, *Physics*, IV. 218^{b}21 ff.

grounds in predicting (or retrodicting) a period of time in which there was no change going on in any part of the universe.

Is it possible to accommodate the possibility of empty time while retaining the insight that time is intimately bound up with change? One way of doing this is to regard times as actual and/or *possible* events. (This is often taken to be what is meant by 'relationism'. It is how the word is used in Le Poidevin's essay. The non-modal counterpart we attributed to Leibniz is sometimes called 'reductionism'.) Another, related, solution, is to regard times as possibilities of temporal location. The way in which a relationist approach to time can take on a modal form is an important topic which has received far less attention than non-modal forms. We hope to go some way towards filling the gap with Graeme Forbes's essay, which has been written especially for this volume.

Despite the name, temporal relationism is not tied to the relational analysis of tensed statements (the B account). It is true that temporal relations, such as the earlier-than relation, occur crucially in the relationist construction of times. But the question is left open whether these relations are ultimately to be understood in tensed terms. For example, on the A account, '*e* is earlier than *f*' is true if and only if: *e* is past and *f* is present or *e* is present and *f* is future or *e* is past and *f* is future or *e* is more past than *f* or *f* is more future than *e*. There is no reason why a believer in tensed facts should not embrace some form of relationism. Indeed, the idea that time just *is* change goes well with the idea of temporal becoming: the flow of time consisting of reality's gradual accretion of more events.

CAUSAL THEORIES OF TIME

The relationist project is an attempt to reduce instants to some other category. But what of temporal *relations*: are these to be reduced to something more fundamental, or are they primitive? If they *are* reducible, what relation could they be reduced to? One very influential answer to this question is that temporal relations are definable in terms of *causal* relations. This central idea gives rise to a number of causal theories of time. On the simplest theory, A is earlier than B if and only if A is among the causes of B, but not vice versa. This rules out the possibility of there being a number of causal series which were causally unconnected to each other, but all sharing a common time-series. (For a version of the causal theory in which this possibility is accommodated, see the Section 'Direction and Possibility', below.)

Causal theories of time rely on the causal relation's being *asymmetric*: that is, if A causes (or is among the causes of) B, then B does not cause (or is not among the causes of) A. Only if the causal relation is asymmetric can the causal theorist explain the direction of time. And, of course, to explain the direction of time is to explain the direction of change. To link this with an earlier debate: we said that those who deny the existence of tensed facts cannot explain the obvious disanalogies between time and space (e.g. the fact that time, not space, has an intrinsic direction) in terms of the passage of time. However, causal theories of time offer an alternative explanation: time has a direction because causation has a direction. We might also add: time is the dimension of change because all changes involve causally connected states of affairs and time is the dimension of causality.

These explanations are totally undermined, however, if it turns out that asymmetry of causation is unrelated to the asymmetry of before and after. Perhaps there are causes which are simultaneous with their effects. It may even be possible for effects to precede their causes, as Michael Dummett argues in Essay VII. Dummett's conclusion is startling, but why do we find it so? Is it merely counter-intuitive, or is its falsity actually entailed by other strongly held beliefs? For instance, the proponent of temporal becoming might provide the following argument:

'A cause makes a difference: it brings about something that would otherwise, in the circumstances, not have happened. Now the natural interpretation of this is that, until the cause occurred, there was no fact of the matter about whether or not the effect would occur. The explanation of the fact that we cannot affect the past lies in the fact that the past is determinate: any statement about the past has a determinate truth-value. It is a fact that p obtained at noon yesterday, so nothing I do now would make a difference to that fact. I can't make it the case that p did not obtain yesterday, so equally I cannot make it the case that p did obtain yesterday. The future, however, being unreal, is *not* determinate in this way. There is no fact of the matter about whether f will obtain tomorrow, so anything I do now *could* make a difference.'

On this explanation of the temporal direction of causation, a directional feature of time, namely temporal becoming, determines the asymmetry of causation, *not* the other way around. If we accept the explanation, then we can make the familiar fatalist objection to those who reject temporal becoming, as follows: 'If there are no ontological differences between the present and the future, then future-tense statements have a determinate truth-value. As there would then be a fact of the matter about what would happen in the future, we could not affect it. Rejection of temporal becoming is therefore a rejection of free will.'

The fatalist argument has been much criticized. It would not be appropriate in this Introduction to come to any conclusions on its cogency, but it is worth

pointing out that the slogan 'a cause makes a difference' need not be interpreted to mean that, prior to the cause, there is simply no fact of the matter
about whether the effect will occur or not. It could mean simply that, if
history had taken a slightly different course from the one it did take, to the
extent that the cause did not occur, then the effect would not have occurred.
Now the truth of this counterfactual seems to be perfectly consistent with the
determinateness of the future. Although it is a fact that q will obtain, if p had
not earlier obtained then it would not be a fact that q will obtain. The debate
does not stop here, however, for one then needs to ask whether the counterfactual relationship is all there is to causation, or whether the counterfactual
relationship only obtains in virtue of there being a causal relationship which
in turn requires temporal becoming.

If we believe that causal theories of time fail, must we accept that the
direction of time is *sui generis*? Not necessarily: we may link temporal
direction with other asymmetries in the universe, such as the increasing
entropy of physical systems. But here, as Lawrence Sklar points out in his
essay, we need to be careful. What is the kind of reduction involved in
explicating temporal direction in terms of entropy? Is it the kind exemplified
in philosophical reductions, such as the phenomenalistic reduction of physi-
cal-object sentences to counterfactual sentences about experiences? Or is it
rather the kind exemplified in scientific identifications, e.g. of light waves
with electromagnetic waves, which need carry no implications for our ordinary, pre-theoretical, understanding of light? If the former, then certain reductionist theories of temporal direction look very implausible. In particular,
Sklar argues, the entropic theory of temporal direction should be seen as a
scientific, not a philosophical, reduction. With Sklar's distinction in mind,
we should reconsider the kind of theory represented by causal theories of
time.

TIME TRAVEL

Although the fact that time has a direction seems constantly forced on our
awareness, there is still room to ask whether it is possible to escape the
constraints it places on us. We can already move freely about the dimensions
of space: is it ridiculous to hope that, one day, we will as freely be able to
move backwards and forwards in the dimension of time? Again, causation
enters the picture, as the possibility of time travel leads to the possibility of
causal loops. The time-traveller steps into her machine, goes back to a past
time, and perhaps there commits certain indiscretions which have causal
consequences which eventually impinge on her years later when she is step-

ping into her time machine: the causal series has come round full circle. Here we have a cause being affected by its own effects, which conflicts with the assumption of asymmetry. Although time-travellers are often exhorted *not* to interfere with history, it is hard to see how they could be thought to *be* at those past (or future) times without interacting with things at those times. But any interaction, however brief and apparently insignificant, will have causal consequences. So any argument against the possibility of causal loops will also be an argument against the possibility of time travel.

Another objection to time travel is open to believers in tensed facts who hold that the present is ontologically privileged. If what is in the past and what is in the future is simply *unreal*, then there is, for the would-be time-traveller, nothing to visit. One might as well try to visit Bleak House.

Most commonly, however, it is argued that time travel makes it possible for the time-traveller so to affect the past that her birth never took place, which has the absurd consequence that one could bring it about that one never existed. But anything which has absurd consequences is itself incoherent, so time travel is not possible. This argument is examined in David Lewis's essay.

CAUSATION AND EMPTY TIME

There are important connections between views about causation and the possibility of time without change. Depending on how one looks at it, one could say either that empty times involve causal anomalies, or that the possibility of empty times requires some revision in our ordinary concept of causation. Consider the situation which is supposed to occur in Shoemaker's story: change altogether ceases for a finite period, and then resumes. What could explain the resumption of change after a changeless period? The state of affairs which immediately preceded the resumption of change was no different from that which obtained immediately after the cessation of change. So why did change resume when it did? There is, it seems, no explanation, unless we look for the explanation in states of affairs obtaining *before* the empty time period. This would mean accepting the possibility of a temporally discontinuous causal chain. So limited periods of empty time appear to face us with a dilemma: either we accept that there can be uncaused events, or there can be action at a temporal distance. We might object that there is a dilemma only if we treat causation as deterministic: that causes are, in the circumstances, sufficient for their effects. However, although allowing in-deterministic causation permits us to talk of a cause of the first event after the empty time period, we would still have to concede that the state of affairs obtaining during this period provides no causal explanation of why that event occurred when it did.

CAUSATION AND TOPOLOGY

Causal theories of time may be used to motivate the ascription of certain topological properties to time. The most obvious example of this concerns the idea, discussed in Essay IX, that time might be closed. In closed time, every event is both before and after every other event. It doesn't follow from this, however, that if a given event is two hours earlier than another it is also two hours later. We are not therefore prevented from characterizing times which are earlier than our present temporal perspective by only a small amount as 'past', and those which are later than our present perspective by only a small amount as 'future'. The idea of closed time is such an abstract and puzzling one that it helps to consider an analogy: because the earth is spherical, London is both to the east and to the west of New York, despite the convention of treating the terms 'the East' and 'the West' as non-relative.

In a closed-time world, as in a time-traveller's world, there could be causal loops. If we add the assumption that the causal relation is transitive, then for any two members of the causal loop, each will be among the causes of the other. As was remarked before, this conflicts with the assumption on which causal theories of time rest: that causation is asymmetric. This alone is enough to make some philosophers pronounce closed time incoherent. Others allow the possibility of closed-time worlds, but deny that time could have a direction in such worlds. One might wonder about the coherence of such a position. After all, what is it that makes a dimension temporal rather than spatial, if not an intrinsic direction? What philosophers are describing under the name of 'closed time' could simply be a fourth spatial dimension. This isn't the end of the story, however, for there remains the interesting possibility that certain conceptions of causation make the relation locally asymmetric. This local asymmetry would impose a local time order. Alternatively, the closed-time proponent could concede that the direction of time was constituted by something other than the causal relation, or insist that causation is asymmetric and non-transitive. Certainly, work needs to be done on the nature of directionality in closed-time worlds before the proponents of closed time could overcome the deeply entrenched intuitions which are antagonistic to it.

Turning now to another topological property: Murray MacBeath's essay challenges our assumptions about the dimensionality of time by showing how certain observations would support the hypothesis that time was two-dimensional. But why do we assume that time is one-dimensional? Are there any a priori reasons in favour of the view? Here again the causal theory of time is relevant. MacBeath defines dimensionality as follows: 'a space has n dimensions if there are n respects in which its occupants can, *qua* occupants of that space, vary continuously but independently'. (Here 'space' is being used in an abstract way which covers colour, temperature, etc.) Consider the space defined by causal relations. How many dimensions will that have? It is not clear how occupants of the space could vary in more than one causal respect: for any two occupants, either one would be either causally anteced-ent to the other, or they would be causally coincident. Causal space, then, is one-dimensional. But if temporal relationships are determined by causal relationships, then time too is one-dimensional. So taking two-dimensional time seriously involves either rejection, or at the very least, radical revision, of causal theories of time. MacBeath's thought experiment confirms this, for the causal relationships between the various events do not appear to impose any constraints on the temporal relationships between those events.

POSSIBLE WORLDS

A third unifying theme for the topics of this volume is that of modality. Modal statements are concerned with what might be the case, what must be the case, or what could not be the case. To put it in different terms, modal statements typically take the form 'Necessarily p' or 'Possibly p'. Other modal statements, exhibiting what Quine calls the third grade of modal involvement,[6] are putatively about purely possible objects. (The discussion here is restricted to so-called 'alethic' modality, as opposed, for example, to 'deontic' modality—what ought to be the case—or epistemic modality—what might (for all we know) be the case.)

Such considerations play no small part in our reflections on ourselves and the world. We consider with regret (or relief) what we might have done. We worry about the possible consequences of our actions. In assigning respons-ibility for actions we must consider what constraints agents were under, and ask ourselves whether some external circumstance necessitated their behav-ing as they did. Such constraints on human behaviour are just cases of constraints on the behaviour of the universe as a whole. We imagine the world to be governed by laws—indeed such a condition appears necessary if we are ever to understand the world. If the world is deterministic, then given the state of the world at any one time only one course of history is compatible

[6] W. V. Quine, 'Three Grades of Modal Involvement', in *The Ways of Paradox* (New York: Random House, 1966), 156–74.

with physical law. The word 'compatible' here introduces the notion of a logical constraint: what could or could not logically be the case given the truth of some proposition. This notion of a logical constraint enters into our view of human rationality. Some beliefs we cannot simultaneously hold without inconsistency. And so on.

In interpreting modal statements, philosophers have made increasing use of the notion of a possible world. An early conception of a possible world, due to Carnap,[7] was that of a maximally consistent set of atomic sentences: for every atomic sentence S, either S or its negation is a member of that set. Other philosophers prefer to talk of sets of propositions, or facts. Informally, a possible world is supposed to be a way this world could have been. With this notion, we are able to interpret 'Necessarily p' as 'p is true in all possible worlds', and 'Possibly p' as 'p is true in at least some world'.

TENSE AND MODALITY

Given the pervasiveness of modal thinking, it is no surprise to find that modality enters into many theories of time. For example, modal notions provide one way of understanding temporal becoming. It has been suggested that the future just is a set of possible worlds. There is no one actual future as there is one actual present (or, for some, one actual past); rather there is only an infinite number of possible futures. Since different states of affairs obtain in these different futures, any statement one makes about the future (other than necessarily true or necessarily false ones) will lack a truth-value.

Something of this is implied in John Lucas's remark that 'It is natural to try and view time as the passage from the possible to the necessary.'[8] However, this remark reflects an older conception of modality. If we think of 'Necessarily p' as 'p is true in all logically possible worlds', then there is no temptation to think of statements about the past as being necessarily true. But we are now used to an atemporal conception of modality, and easily forget that such a conception is a relatively recent development. In Aristotle, we find a modality that is explicitly relativized to a time. In *De Caelo*, for example, according to various commentators,[9] the following principles are operating: (1) If it is possible that p, then at some time it is the case that p; (2) If it is always the case that p, then it is necessary that p. This has the consequence that statements may change their modal properties over time,

---

[7] Rudolf Carnap, *Meaning and Necessity* (Chicago: University of Chicago Press, 1947).

[8] J. R. Lucas, *A Treatise on Time and Space* (London: Methuen, 1973), 262.

[9] See e.g. Jaako Hintikka, *Time and Necessity* (Oxford: Clarendon Press, 1967); Sarah Waterlow, *Passage and Possibility* (Oxford: Clarendon Press, 1982).

hence Lucas's observation. What was the case is necessarily so in the sense that we cannot change it. In contrast, talk of what will be the case is often replaced by talk of what might be the case.

The links between the development of a logic of possibility and a logic of time have been very close. To begin with, there are clear structural parallels between modal logic and Prior's tense logic:

1. Tensed terms are treated as sentential operators: 'It was the case that' (compare 'It is possibly the case that').
2. The tense-logical axioms of various systems to some extent mirror modal axioms.
3. Tensed propositions, such as 'There is a thunderstorm going on', are treated as complete propositions which may change their truth-value over time. It is not necessary to add a reference to a time in order to complete the proposition. (Similarly, 'I am (actually) six foot' is a complete proposition which has a different truth-value in different worlds.)

The connection between tense logic and modal logic may be more intimate than this, however. Indeed one has been seen as a branch of the other. J. N. Findlay wrote in 1941: 'the calculus of tenses should have been included in the modern development of modal logics'.[10] It is not clear whether the implication is that the temporal expressions should be interpreted in modal terms or rather that the modal expressions should be interpreted in temporal terms. Prior explored the second way, and discovered that certain axioms in modal logic are intuitively plausible (and indeed provable) if 'Possibly p' is defined as 'Either it is the case that p or it will be the case that p'.

The treatment of temporal logic as a logic of tenses, though widespread, is not universally endorsed. Rescher and Urquhart, for example, prefer a quantified temporal logic,[11] in which, instead of tense operators, there are quantifiers ranging over instants and intervals, and terms standing for temporal relations. This treatment of the logic of time is entirely appropriate for those who reject tensed facts.

DIRECTION AND POSSIBILITY

Even if time is not the passage from the possible to the necessary, modality may enter into a different account of the direction of time, namely, that provided by causal theories of time. The causal theorist may want to allow

[10] J. N. Findlay, 'Time: A Treatment of Some Puzzles', *Australasian Journal of Philosophy*, 19 (1941): 216–35.
[11] Nicholas Rescher and Alasdair Urquhart, *Temporal Logic* (New York: Springer-Verlag, 1971).

for the possibility of a temporally unified but causally disunified universe. If so, then a natural amendment of the causal analysis of time order could be constructed along the following lines: 'e is earlier than $e*$ if and only if it is possible for there to be a causal series running from e to $e*$ such that e is causally antecedent to states which are themselves causally antecedent to $e*$'. The sense of 'possible' needs to be qualified if the account is to be a plausible one, for in the widest sense of the term it is possible for any two events (including simultaneous ones) to be so related that one is causally antecedent to the other. This point depends not on the logical possibility of simultaneous causation, but rather on the possibility of events standing in different temporal relations to each other from those in which they actually stand. Since it is possible (in the wide sense of not involving a logical contradiction) for two events which are actually simultaneous to be, in some other world, non-simultaneous, it is possible for one to be causally antecedent to the other. This clearly would undermine the causal account. The kind of possibility relevant to causal analyses of time, obviously, cannot be mere logical possibility. In some causal theories of time (for example, Reichenbach's[12]) the modality is a physical one: e is earlier than e' if and only if it is physically possible for there to be a causal series running from e to e'. Here 'physically possible' might simply be interpreted as 'compatible with physical law', but that hardly seems adequate. After all, it is presumably compatible with physical law that the temporal relation between two token events could be reversed. What is needed is something like this: e is earlier than e' if and only if, in the circumstances, it is physically possible . . . etc. In other words, the kinds of restriction are those appropriate for the interpretation of counterfactual conditionals.[13]

The introduction of modality into the causal theory of time, however, raises difficulties. One kind of difficulty is epistemological. How do we come to be aware of temporal relations? In the simplest case, we may see one event leading to another. Here we are aware of temporal relations through witnessing causal relations. Now if causality is the mechanism by which we have knowledge of temporal relations, then there is some epistemological motivation for a non-modal causal theory of time: one which constructs temporal relations out of actual causal relations. There is no such motivation for the modal version. We do not, surely, become aware of temporal relations through being aware of the possibility of causal relations; it is, if anything, the other way around. Another kind of difficulty with the introduction of

[12] H. Reichenbach, *Axiomatik der relativitischen Raum-Zeit Lehre* (Braunschweig: Viewig, 1924), 22.
[13] See e.g. David Lewis, *Counterfactuals* (Oxford: Blackwell, 1973).

modality is ontological. What is it that grounds the possibility of causal relations between events if not the actual temporal relations between them?

TIMES AS POSSIBLE EVENTS

Similar epistemological and ontological worries can be raised for relationism. The relationist takes times to be collections of actual and possible events (or, perhaps not equivalently, as possibilities of temporal location). Now while it is plausible to say that we are aware of the existence of times in virtue of being aware of the various events that occur at those times, it is not at all plausible to say that we are aware of the existence of times in virtue of being aware of the possibility of change n units after/before some given event. And (the related ontological worry) it is only because there exists a time between two events that there is the possibility of change between those events.

In order to address these kinds of difficulty, proponents of modal theories of time need to consider what account of modality best fits their own theory of time. Modality, like time, is a very obscure notion. It is not immediately apparent that, by defining the second in terms of the first, we have made a big step forward in clarifying our ideas. What, after all, are the truth-conditions of modal statements? Out of what are possibilia constructed? Some answers to these questions may require one to take times as given, as 'ontologically basic', but this need not be true of all of them. There has been a great deal of important work on modality in recent years, and we are now in a much better position than we were to see just how modal theories of time interact with different theories of modality.

MODALITY AND TOPOLOGY

Finally, what is the connection between modality and questions concerning the topology of time? Is it, for example, a necessary truth (or falsehood) that time has neither a beginning nor an end? Or is this only contingently true (or false)? And what of the proposition that between any two moments there is always a third, or that all actual events are temporally related to each other? There is a long tradition, dating back to Aristotle, of taking statements of this kind—statements concerning the topological structure of time—to be necessary.

This view has been supported by a number of a priori arguments in favour of or against certain topologies of time. If such arguments are successful, then it is an essential feature of time that it has such and such topological properties. However, these a priori arguments have sometimes pulled in opposite directions. For example, Aristotle argued that one could go on dividing a temporal interval without ever reaching the limit of the process.[14] Hume, in contrast, argued that, since in our perception of time there are experiential minima (in other words that an interval of less than a certain finite duration would not be perceived as an interval), time itself must be composed of discrete instants.[15] Perhaps the most famous examples of such conflicting arguments are found in the work of Kant, who deliberately exploited them to motivate his view that time is not an aspect of an external reality, but rather a scheme we impose upon our experience in order to be able to interpret it. In the *Critique of Pure Reason* Kant sets a compelling argument to the effect that time had a beginning side by side with an equally compelling argument to the effect that time had no beginning. Perhaps Kant's antimonies are the source of the feeling that a priori arguments for time's having a certain topological structure are doomed to end in failure. This feeling has certainly grown in our own times. Bill Newton-Smith's *The Structure of Time* is in part an attack on these arguments, and one of the main conclusions of the book is that the topology of time is a matter to be settled empirically. In his contribution to this anthology, Newton-Smith tackles the question of whether or not time is bounded. Although he leaves the ultimate solution to the problem to the physicists, his essay demonstrates the important role philosophers have to play in the debate.

In the course of his essay, Newton-Smith comments on a certain kind of argument for time's unboundedness, which goes as follows: Suppose E to be the first event. This is a purely contingent fact about E, for it is possible for there to have been events earlier than E. But for there to be this possibility, there must actually be times before E. One can repeat the argument for any putative first moment, so as to demonstrate that there cannot be a first moment to time. Newton-Smith finds the argument unconvincing. It can be made to look far more plausible, however, if one adopts a relationist view, in which there is a conceptual connection between the possibility of change before E and the actual existence of times before E. Moreover, this kind of argument can be used to provide a priori motivation for other views about the topology of time, for example, that time is continuous and linear. The connections between relationism and topology are explored in Le Poidevin's essay. The essay as originally published is followed by a Post-script in which Le Poidevin attacks some of his earlier arguments and conclusions.

[14] Aristotle, *Physics*, VI, esp. 232^{b}20 ff.
[15] David Hume, *A Treatise of Human Nature*, ed. L. A. Selby-Bigge, rev. P. H. Nidditch (Oxford: Clarendon Press, 1978), 26–33.

FACT AND FANTASY

Le Poidevin's essay is concerned with the problem to what extent questions about the nature of time are to be answered by empirical observation, experiment, and theory, to what extent by armchair, or a priori, argumentation. It is a feature of this anthology that the reader of it should be able to understand and assess nearly all the essays in it in virtually complete ignorance of, say, relativity theory, which has revolutionized our understanding of time in the twentieth century. Not all that comes under the heading of the philosophy of time is like this: much of the work done by philosophers of time consists in an examination of the nature of the arguments for, or of the implications of, scientific theories such as are associated with the names of Albert Einstein and, more recently, Stephen Hawking. The philosophy of time that is practised in that mode we regard as a branch of the philosophy of science, whereas this anthology is devoted to the philosophy of time in its metaphysical mode. Metaphysics we defined, at the beginning of this Introduction, as 'the philosophical study of what there is'. If the metaphysician is interested in what there *is* or in the way things *are*, it is natural that he should raise the question whether things are the way things are because that is the way things *must* be (and we have discussed the central role of modal notions in 'Possible Worlds', above). It is also natural that, if one wants to question the idea that, in this respect or that, things must be the way they are, one should try to give an account of what it would be for things to be otherwise. This account will be a story, a (partial) description of an allegedly possible world; when the claim is added that the coherence of the story establishes the logical (or conceptual) possibility of the feature central to the story, we have a 'fantasy argument'. Mellor, who speaks disparagingly of 'this fashionable kind of argument by fantasy',[16] describes it thus: it 'presumes to show something possible by describing an imaginary world in which we should apparently be inclined to believe the possibility actual'.[17]

Mellor's reference to belief is important: fantasy arguments usually proceed not just by describing an allegedly possible world, and claiming that this is a world in which time is, say, disunified or two-dimensional; rather, they seek to portray a world which is such that its inhabitants have reason (of a respectable empirical kind) to believe its time system to be disunified or two-dimensional. In this anthology Shoemaker attempts to portray a

---

[16] D. H. Mellor, 'Theoretically Structured Time', *Philosophical Books*, 23 (1982): 65–9: 66. When Mellor says 'this . . . kind' he is referring to an argument of Bill Newton-Smith's in his *The Structure of Time*. See further W. H. Newton-Smith: 'Reply to Dr Mellor' (ibid. 69–71).
[17] Mellor, 'Theoretically Structured Time', 66

world which, its inhabitants have good reason to believe, undergoes stretches of changeless time. As we said earlier, Shoemaker has to deal with Aristotle's observation that changeless time cannot be experienced; and his argument embodies two crucial elements: first, an appeal to the testimony of others, based on their experience; and secondly, an invocation of the process of theoretical extrapolation. Shoemaker's people are not directly aware of temporal 'freezes'—indeed, Shoemaker implies that Aristotle was right to reject that possibility—but they do have indirect evidence of the existence of such freezes. MacBeath too, in his fantasy argument, is careful to argue not just that a two-dimensional-time world is conceivable, from the outside, as it were, but that its inhabitants could have reason to think of time in their world as being two-dimensional; and he too argues that a large part of that reason would be supplied by the testimony of others.

Anthony Quinton, in Essay XII, puts forward a fantasy argument designed to establish the possibility of disunified space. He thinks, however, that no parallel argument can be constructed for disunified time, because evidence for the existence of two temporally unrelated time systems would have to be two experiences in a single temporal sequence. What Shoemaker did for Aristotle, Richard Swinburne did for Quinton in an article, which he later repudiated, with the famous fantasy of the Okku and the Bokku.[18] Here again a crucial part of the case for saying that the Bokku have reason to believe that there are two temporally unrelated time systems in their world is the testimony of the Okku.

Are the fantasy arguments of MacBeath, Quinton, Shoemaker, and Swinburne anything more than ingenious stories, with, as stories, no probative force whatsoever? Shoemaker's argument is undoubtedly ingenious, not least in that it provides a way of answering the question how we could assign a particular length to a period of changeless time, during which, *ex hypothesi*, anything that might be taken as a clock would be stopped (though what a year is in his world he does not tell us). How well does his argument conform to Mellor's description of a fantasy argument as one which 'presumes to show something possible by describing an imaginary world in which we should apparently be inclined to believe the possibility actual'? That depends on how we construe the phrase 'in which we should apparently be inclined': does it mean 'such that, if we inhabited it, we would (for so the fantasy story goes) be inclined'? If so, Mellor is quite right to dismiss arguments by fantasy. An answer 'I would say that *p*' to the question 'What would you say if this, that, and the other happened?' tells us nothing about the possibility, logical or otherwise, of *p*, as is shown by the following exchange. *Adam*:

---

[18] For references to the original fantasy and the recantation, see Essay XI nn. 1–3.

'What would you say if, while you were under post-hypnotic suggestion, the hypnotist gave the cue for you to say "There is a prime number greater than two which is even"?' *Eve*: 'I would say that there is a prime number greater than two which is even.'

But there is another interpretation of Mellor's phrase 'in which we should apparently be inclined': on this interpretation we, as inhabitants of our world, are being asked, in effect, what we, here and now, think should (or ought to) be said about the allegedly possible world. Ought we to agree with the inhabitants of Shoemaker's world, who, on the evidence available to them, think they have good reason to postulate a year every sixty years during which no change occurs anywhere in their world? Here we engage with Mellor's objection to fantasy arguments: 'to show possibility as well as sense, one has also to show the imaginary world possible, which a merely plausible sketch of it does not do. Impossibilities are all too easy to make plausible. . . . In the face of contrary arguments *a priori* . . . fantasies supply no evidence for possibility at all.'[19]

It is not clear how one 'shows an imaginary world possible' other than by discrediting arguments for its impossibility; and, in the absence of any guidance on this, it is hard to see how to apply Mellor's argument that one should show the world to be possible before one concludes that this or that hypothesis, which is made to look plausible in the world in question, could be true. Fantasy arguments do at least have the virtue that those who deny that this or that (backwards causation, changeless, disunified, or two-dimensional time) is possible are forced not to reject the possibility out of hand: even if they are right to reject the possibility in question, the impossibility may lie less near the surface than first appears.

Besides the philosophy of space and time, another area of philosophy in which the use of fantasy arguments has been very common is the discussion of personal identity.[20] Kathy Wilkes has recently argued that the philosophical discussion of personal identity can and should proceed, to quote from the subtitle of her book *Real People*,[21] 'without thought experiments'. One of her arguments is that, in this sphere, the truth is often stranger than fiction. Very much the same could be said of time. Indeed, if one considers the familiar story of the twin who, after a return trip into space at near the speed of light, finds that she is much younger than her earthbound sister, one might

---

[19] Mellor, 'Theoretically Structured Time', 66. Peter Geach, in a similar vein, speaks of the 'trap of producing in [oneself] a spurious understanding of a really incoherent story—a trap . . . that yawns for all who use fantasy in philosophy' ('Reincarnation', in his *God and the Soul* (London: Routledge & Kegan Paul, 1969), 14).

[20] See J. L. H. Thomas, 'Against the Fantasts', *Philosophy*, 66 (1991): 349–67, 356.

[21] Kathleen V. Wilkes, *Real People: Personal Identity without Thought Experiments* (Oxford: Clarendon Press, 1988).

well wonder whether we are in the realm of fact or of fantasy. Of fact, presumably, in that the differential ageing of the twins is a result entailed by an empirically well-supported scientific theory; but the experiment as described will only ever be conducted in thought. Thought experiments abound both in science and in the philosophy of science. In this anthology, Sklar's essay is the one which comes closest to being philosophy of time done in the philosophy-of-science mode; his essay does presuppose acquaintance with some notions from twentieth-century science; and it is perhaps the essay least touched by the spirit of fantasy. It is striking, therefore, that it should be this essay which contains the engagingly artless sentence 'Suppose . . . that some fairly substantial miracles occurred in this world.'

1

TIME AND TENSE

I

THE UNREALITY OF TIME

J. M. E. McTAGGART

It will be convenient to begin our enquiry by asking whether anything exist-ent can possess the characteristic of being in time. I shall endeavour to prove that it cannot.

It seems highly paradoxical to assert that time is unreal, and that all statements which involve its reality are erroneous. Such an assertion in-volves a departure from the natural position of mankind which is far greater than that involved in the assertion of the unreality of space or the unreality of matter. For in each man's experience there is a part—his own states as known to him by introspection—which does not even appear to be spa-tial or material. But we have no experience which does not appear to be temporal. Even our judgements that time is unreal appear to be themselves in time.

Yet in all ages and in all parts of the world the belief in the unreality of time has shown itself to be singularly persistent. In the philosophy and religion of the West—and still more, I suppose, in the philosophy and reli-gion of the East—we find that the doctrine of the unreality of time contin-ually recurs. Neither philosophy nor religion ever hold themselves apart from mysticism for any long period, and almost all mysticism denies the reality of time. In philosophy, time is treated as unreal by Spinoza, by Kant, and by Hegel. Among more modern thinkers, the same view is taken by Mr Bradley. Such a concurrence of opinion is highly significant, and is not the less significant because the doctrine takes such different forms, and is supported by such different arguments.

I believe that nothing that exists can be temporal, and that therefore time is unreal. But I believe it for reasons which are not put forward by any of the philosophers I have just mentioned.

This essay is drawn from chapter 33 of J. M. E. McTaggart's *The Nature of Existence*, ii (Cambridge: Cambridge University Press, 1927). The chapter is entitled simply 'Time'; but, because it is a restatement of arguments McTaggart had advanced in his article 'The Unreality of Time' (*Mind*, 17 (1908): 457–74), the editors of this volume think it not inappropriate to give this essay the longer, more descriptive title. Reprinted by permission of Cambridge University Press.

Positions in time, as time appears to us prima facie, are distinguished in two ways. Each position is Earlier than some and Later than some of the other positions. To constitute such a series there is required a transitive asymmetrical relation, and a collection of terms such that, of any two of them, either the first is in this relation to the second, or the second is in this relation to the first. We may take here either the relation of 'earlier than' or the relation of 'later than', both of which, of course, are transitive and asymmetrical. If we take the first, then the terms have to be such that, of any two of them, either the first is earlier than the second, or the second is earlier than the first.

In the second place, each position is either Past, Present, or Future. The distinctions of the former class are permanent, while those of the latter are not. If M is ever earlier than N, it is always earlier. But an event, which is now present, was future, and will be past.

Since distinctions of the first class are permanent, it might be thought that they were more objective, and more essential to the nature of time, than those of the second class. I believe, however, that this would be a mistake, and that the distinction of past, present, and future is as *essential* to time as the distinction of earlier and later, while in a certain sense it may . . . be regarded as more *fundamental* than the distinction of earlier and later. And it is because the distinctions of past, present, and future seem to me to be essential for time that I regard time as unreal.

For the sake of brevity I shall give the name of the A series to that series of positions which runs from the far past through the near past to the present, and then from the present through the near future to the far future, or conversely. The series of positions which runs from earlier to later, or conversely, I shall call the B series. The contents of any position in time form an event. The varied simultaneous contents of a single position are, of course, a plurality of events. But, like any other substance, they form a group, and this group is a compound substance. And a compound substance consisting of simultaneous events may properly be spoken of as itself an event.[1]

---

[1] It is very usual to contemplate time by the help of a metaphor of spatial movement. But spatial movement in which direction? The movement of time consists in the fact that later and later terms pass into the present, or—which is the same fact expressed in another way—that presentness passes to later and later terms. If we take it the first way, we are taking the B series as sliding along a fixed A series. If we take it the second way, we are taking the A series as sliding along a fixed B series. In the first case time presents itself as a movement from future to past. In the second case it presents itself as a movement from earlier to later. And this explains why we say that events come out of the future, while we say that we ourselves move towards the future. For each man identifies himself especially with his present state, as against his future or his past, since it is the only one which he is directly perceiving. And this leads him to say that he is moving with the present towards later events. And as those events are now future, he says that he is moving towards the future.

Thus the question as to the movement of time is ambiguous. But if we ask what is the movement of either series, the question is not ambiguous. The movement of the A series along the B series is from earlier to later. The movement of the B series along the A series is from future to past.

The first question which we must consider is whether it is essential to the reality of time that its events should form an *A* series as well as a *B* series. It is clear, to begin with, that, in present experience, we never *observe* events in time except as forming both these series. We perceive events in time as being present, and those are the only events which we actually perceive. And all other events which, by memory or by inference, we believe to be real, we regard as present, past, or future. Thus the events of time as observed by us form an *A* series.

It might be said, however, that this is merely subjective. It might be the case that the distinction of positions in time into past, present, and future is only a constant illusion of our minds, and that the real nature of time contains only the distinctions of the *B* series—the distinctions of earlier and later. In that case we should not perceive time as it really is, though we might be able to *think* of it as it really is.

This is not a very common view, but it requires careful consideration. I believe it to be untenable, because, as I said above, it seems to me that the *A* series is essential to the nature of time, and that any difficulty in the way of regarding the *A* series as real is equally a difficulty in the way of regarding time as real.

It would, I suppose, be universally admitted that time involves change. In ordinary language, indeed, we say that something can remain unchanged through time. But there could be no time if nothing changed. And if anything changes, then all other things change with it. For its change must change some of their relations to it, and so their relational qualities. The fall of a sand-castle on the English coast changes the nature of the Great Pyramid.

If, then, a *B* series without an *A* series can constitute time, change must be possible without an *A* series. Let us suppose that the distinctions of past, present, and future do not apply to reality. In that case, can change apply to reality?

What, on this supposition, could it be that changes? Can we say that, in a time which formed a *B* series but not an *A* series, the change consisted in the fact that the event ceased to be an event, while another event began to be an event? If this were the case, we should certainly have got a change.

But this is impossible. If *N* is ever earlier than *O* and later than *M*, it will always be, and has always been, earlier than *O* and later than *M*, since the relations of earlier and later are permanent. *N* will thus always be in a *B* series. And as, by our present hypothesis, a *B* series by itself constitutes time, *N* will always have a position in a time-series, and always has had one. That is, it always has been an event, and always will be one, and cannot begin or cease to be an event.

Or shall we say that one event M merges itself into another event N, while still preserving a certain identity by means of an unchanged element, so that it can be said, not merely that M has ceased and N begun, but that it is M which has become N? Still the same difficulty recurs. M and N may have a common element, but they are not the same event, or there would be no change. If, therefore, M changed into N at a certain moment, then, at that moment, M would have ceased to be M, and N would have begun to be N. This involves that, at that moment, M would have ceased to be an event, and N would have begun to be an event. And we saw, in the last paragraph, that, on our present hypothesis, this is impossible.

Nor can such change be looked for in the different moments of absolute time, even if such moments should exist. For the same argument will apply here. Each such moment will have its own place in the B series, since each would be earlier or later than each of the others. And, as the B series depends on permanent relations, no moment could ever cease to be, nor could it become another moment.

Change, then, cannot arise from an event ceasing to be an event, nor from one event changing into another. In what other way can it arise? If the characteristics of an event change, then there is certainly change. But what characteristics of an event can change? It seems to me that there is only one class of such characteristics. And that class consists of the determinations of the event in question by the terms of the A series.

Take any event—the death of Queen Anne, for example—and consider what changes can take place in its characteristics. That it is a death, that it is the death of Anne Stuart, that it has such causes, that it has such effects— every characteristic of this sort never changes. 'Before the stars saw one another plain', the event in question was the death of a queen. At the last moment of time—if time has a last moment—it will still be the death of a queen. And in every respect but one, it is equally devoid of change. But in one respect it does change. It was once an event in the far future. It became every moment an event in the nearer future. At last it was present. Then it became past, and will always remain past, though every moment it becomes further and further past.[2]

Such characteristics as these are the only characteristics which can change. And, therefore, if there is any change, it must be looked for in the A series, and in the A series alone. If there is no real A series, there is no real change.

---

[2] The past, therefore, is always changing, if the A series is real at all, since at each moment a past event is further in the past than it was before. This result follows from the reality of the A series, and is independent of the truth of our view that all change depends exclusively on the A series. It is worth while to notice this, since most people combine the view that the A series is real with the view that the past cannot change—a combination which is inconsistent.

The *B* series, therefore, is not by itself sufficient to constitute time, since time involves change.

The *B* series, however, cannot exist except as temporal, since earlier and later, which are the relations which connect its terms, are clearly time-relations. So it follows that there can be no *B* series when there is no *A* series, since without an *A* series there is no time.

We must now consider three objections which have been made to this position. The first is involved in the view of time which has been taken by Mr Russell, according to which past, present, and future do not belong to time *per se*, but only in relation to a knowing subject. An assertion that *N* is present means that it is simultaneous with that assertion, an assertion that it is past or future means that it is earlier or later than that assertion. Thus it is only past, present, or future in relation to some assertion. If there were no consciousness, there would be events which were earlier and later than others, but nothing would be in any sense past, present, or future. And if there were events earlier than any consciousness, those events would never be future or present, though they could be past.

If *N* were ever present, past, or future in relation to some assertion *V*, it would always be so, since whatever is ever simultaneous to, earlier than, or later than *V* will always be so. What, then, is change? We find Mr Russell's views on this subject in his *Principles of Mathematics*, section 442. 'Change is the difference, in respect of truth or falsehood, between a proposition concerning an entity and the time *T*, and a proposition concerning the same entity and the time *T'*, provided that these propositions differ only by the fact that *T* occurs in the one where *T'* occurs in the other.' That is to say, there is change, on Mr Russell's view, if the proposition 'At the time *T* my poker is hot' is true, and the proposition 'At the time *T'* my poker is hot' is false.

I am unable to agree with Mr Russell. I should, indeed, admit that, when two such propsitions were respectively true and false, there would be change. But then I maintain that there can be no time without an *A* series. If, with Mr Russell, we reject the *A* series, it seems to me that change goes with it, and that therefore time, for which change is essential, goes too. In other words, if the *A* series is rejected, no proposition of the type 'At the time *T* my poker is hot' can ever be true, because there would be no time.

It will be noticed that Mr Russell looks for change, not in the events in the time-series, but in the entity to which those events happen, or of which they are states. If my poker, for example, is hot on a particular Monday, and never before or since, the event of the poker being hot does not change. But the poker changes, because there is a time when this event is happening to it, and a time when it is not happening to it.

But this makes no change in the qualities of the poker. It is always a quality of that poker that it is one which is hot on that particular Monday. And it is always a quality of that poker that it is one which is not hot at any other time. Both these qualities are true of it at any time—the time when it is hot and the time when it is cold. And therefore it seems to be erroneous to say that there is any change in the poker. The fact that it is hot at one point in a series and cold at other points cannot give change, if neither of these facts change—and neither of them does. Nor does any other fact about the poker change, unless its presentness, pastness, or futurity change.

Let us consider the case of another sort of series. The meridian of Greenwich passes through a series of degrees of latitude. And we can find two points in this series, S and S', such that the proposition 'At S the meridian of Greenwich is within the United Kingdom' is true, while the proposition 'At S' the meridian of Greenwich is within the United Kingdom' is false. But no one would say that this gave us change. Why should we say so in the case of the other series?

Of course there is a satisfactory answer to this question if we are correct in speaking of the other series as a time-series. For where there is time, there is change. But then the whole question is whether it is a time-series. My contention is that if we remove the A series from the prima facie nature of time, we are left with a series which is not temporal, and which allows change no more than the series of latitudes does.

If, as I have maintained, there can be no change unless facts change, then there can be no change without an A series. For, as we saw with the death of Queen Anne, and also in the case of the poker, no fact about anything can change, unless it is a fact about its place in the A series. Whatever other qualities it has, it has always. But that which is future will not always be future, and that which was past was not always past.

It follows from what we have said that there can be no change unless some propositions are sometimes true and sometimes false. This is the case of propositions which deal with the place of anything in the A series—'The Battle of Waterloo is in the past', 'It is now raining'. But it is not the case with any other propositions.

Mr Russell holds that such propositions are ambiguous, and that to make them definite we must substitute propositions which are always true or always false—'The Battle of Waterloo is earlier than this judgement', 'The fall of rain is simultaneous with this judgement'. If he is right, all judgements are either always true, or always false. Then, I maintain, no facts change. And then, I maintain, there is no change at all.

I hold, as Mr Russell does, that there is no A series. (My reasons for this will be given below, pp. 31–4.) And . . . I regard the reality lying behind the appearance of the *A* series in a manner not completely unlike that which Mr Russell has adopted. The difference between us is that he thinks that, when the *A* series is rejected, change, time, and the *B* series can still be kept, while I maintain that its rejection involves the rejection of change, and, consequently, of time, and of the *B* series.

The second objection rests on the possibility of non-existent time-series—such, for example, as the adventures of Don Quixote. This series, it is said, does not form part of the *A* series. I cannot at this moment judge it to be either past, present, or future. Indeed, I know that it is none of the three. Yet, it is said, it is certainly a *B* series. The adventure of the galley-slaves, for example, is later than the adventure of the windmills. And a *B* series involves time. The conclusion drawn is that an *A* series is not essential to time.

I should reply to this objection as follows. Time only belongs to the existent. If any reality is in time, that involves that the reality in question exists. This, I think, would be universally admitted. It may be questioned whether all of what exists is in time, or even whether anything really existent is in time, but it would not be denied that, if anything is in time, it must exist.

Now what is existent in the adventures of Don Quixote? Nothing. For the story is imaginary. The states of Cervantes' mind when he invented the story, the states of my mind when I think of the story—these exist. But then these form part of an *A* series. Cervantes' invention of the story is in the past. My thought of the story is in the past, the present, and—I trust—the future.

But the adventures of Don Quixote may be believed by a child to be historical. And in reading them I may, by an effort of my imagination, contemplate them as if they really happened. In this case, the adventures are believed to be existent, or are contemplated as existent. But then they are believed to be in the *A* series, or are contemplated as being in the *A* series. The child who believes them to be historical will believe that they happened in the past. If I contemplate them as existent, I shall contemplate them as happening in the past. In the same way, if I believed the events described in Jefferies' *After London* to exist, or contemplated them as existent, I should believe them to exist in the future, or contemplate them as existing in the future. Whether we place the object of our belief or of our contemplation in the present, the past, or the future will depend upon the characteristics of that object. But somewhere in the *A* series it will be placed.

Thus the answer to the objection is that, just as far as a thing is in time, it is in the *A* series. If it is really in time, it is really in the *A* series. If it is believed to be in time, it is believed to be in the *A* series. If it is contemplated as being in time, it is contemplated as being in the *A* series.

The third objection is based on the possibility that, if time were real at all, there might be in reality several real and independent time-series. The objection, if I understand it rightly, is that every time-series would be real, while the distinctions of past, present, and future would only have a meaning within each series, and would not, therefore, be taken as absolutely real. There would be, for example, many presents. Now, of course, many points of time can be present. In each time-series many points are present, but they must be present successively. And the presents of the different time-series would not be successive, since they are not in the same time.[3] And different presents, it would be said, cannot be real unless they are successive. So the different time-series, which are real, must be able to exist independently of the distinction between past, present, and future.

I cannot, however, regard this objection as valid. No doubt in such a case, no present would be *the* present—it would only be the present of a certain aspect of the universe. But then no time would be *the* time—it would only be the time of a certain aspect of the universe. It would be a real time-series, but I do not see that the present would be less real than the time.

I am not, of course, maintaining that there is no difficulty in the existence of several distinct A series. In the second part of this chapter I shall endeavour to show that the existence of *any* A series is impossible. What I assert here is that, if there could be an A series at all, and if there were any reason to suppose that there were several distinct B series, there would be no additional difficulty in supposing that there should be a distinct A series for each B series.

We conclude, then, that the distinctions of past, present, and future are essential to time, and that, if the distinctions are never true of reality, then no reality is in time. This view, whether true or false, has nothing surprising in it. It was pointed out above that we always perceive time as having these distinctions. And it has generally been held that their connection with time is a real characteristic of time, and not an illusion due to the way in which we perceive it. Most philosophers, whether they did or did not believe time to be true of reality, have regarded the distinctions of the A series as essential to time.

When the opposite view has been maintained it has generally been, I believe, because it was held (rightly, as I shall try to show) that the distinctions of past, present, and future cannot be true of reality, and that consequently, if the reality of time is to be saved, the distinction in question must be shown to be unessential to time. The presumption, it was held, was for the reality of time, and this would give us a reason for rejecting the A series as unessential to time. But, of course, this could only give a presumption. If

---

[3] Neither would they be simultaneous, since that equally involves being in the same time. They would stand in no time-relation to one another.

the analysis of the nature of time has shown that, by removing the A series, time is destroyed, this line of argument is no longer open.

I now pass to the second part of my task. Having, as it seems to me, succeeded in proving that there can be no time without an A series, it remains to prove that an A series cannot exist, and that therefore time cannot exist. This would involve that time is not real at all, since it is admitted that the only way in which time can be real is by existing.

Past, present, and future are characteristics which we ascribe to events, and also to moments of time, if these are taken as separate realities. What do we mean by past, present, and future? In the first place, are they relations or qualities? It seems quite clear to me that they are not qualities but relations, though, of course, like other relations, they will generate relational qualities in each of their terms.[4] But even if this view should be wrong, and they should in reality be qualities and not relations, it will not affect the result which we shall reach. For the reasons for rejecting the reality of past, present, and future, which we are about to consider, would apply to qualities as much as to relations.

If, then, anything is to be rightly called past, present, or future, it must be because it is in relation to something else. And this something else to which it is in relation must be something outside the time-series. For the relations of the A series are changing relations, and no relations which are exclusively between members of the time-series can ever change. Two events are exactly in the same places in the time-series, relatively to one another, a million years before they take place, while each of them is taking place, and when they are a million years in the past. The same is true of the relation of moments to one another, if moments are taken as separate realities. And the same would be true of the relations of events to moments. The changing relation must be to something which is not in the time-series.

Past, present, and future, then, are relations in which events stand to something outside the time-series. Are these relations simple, or can they be defined? I think that they are clearly simple and indefinable. But, on the other hand, I do not think that they are isolated and independent. It does not seem that we can know, for example, the meaning of pastness, if we do not know the meaning of presentness or of futurity.

---

[4] It is true, no doubt, that my anticipation of an experience M, the experience itself, and the memory of the experience are three states which have different original qualities. But it is not the future M, the present M, and the past M which have these three different qualities. The qualities are possessed by three different events—the anticipation of M, M itself, and the memory of M—each of which in its turn is future, present, and past. Thus this gives no support to the view that the changes of the A series are changes of original qualities.

We must begin with the *A* series, rather than with past, present, and future, as separate terms. And we must say that a series is an *A* series when each of its terms has, to an entity *X* outside the series, one, and only one, of three indefinable relations, pastness, presentness, and futurity, which are such that all the terms which have the relation of presentness to *X* fall between all the terms which have the relation of pastness to *X*, on the one hand, and all the terms which have the relation of futurity to *X*, on the other hand.

We have come to the conclusion that an *A* series depends on relations to a term outside the *A* series. This term, then, could not itself be in time, and yet must be such that different relations to it determine the other terms of those relations, as being past, present, or future. To find such a term would not be easy, and yet such a term must be found, if the *A* series is to be real. But there is a more positive difficulty in the way of the reality of the *A* series.

Past, present, and future are incompatible determinations. Every event must be one or the other, but no event can be more than one. If I say that any event is past, that implies that it is neither present nor future, and so with the others. And this exclusiveness is essential to change, and therefore to time. For the only change we can get is from future to present, and from present to past.

The characteristics, therefore, are incompatible. But every event has them all.[5] If *M* is past, it has been present and future. If it is future, it will be present and past. If it is present, it has been future and will be past. Thus all the three characteristics belong to each event. How is this consistent with their being incompatible?

It may seem that this can easily be explained. Indeed, it has been impossible to state the difficulty without almost giving the explanation, since our language has verb-forms for the past, present, and future, but no form that is common to all three. It is never true, the answer will run, that *M is* present, past, and future. It *is* present, *will be* past, and *has been* future. Or it *is* past, and *has been* future and present, or again *is* future, and *will be* present and past. The characteristics are only incompatible when they are simultaneous, and there is no contradiction to this in the fact that each term has all of them successively.

But what is meant by 'has been' and 'will be'? And what is meant by 'is', when, as here, it is used with a temporal meaning, and not simply for predication? When we say that *X* has been *Y*, we are asserting *X* to be *Y* at a moment of past time. When we say that *X* will be *Y*, we are asserting *X* to be

---

[5] If the time-series has a first term, that term will never be future, and if it has a last term, that term will never be past. But the first term, in that case, will be present and past, and the last term will be future and present. And the possession of two incompatible characteristics raises the same difficulty as the possession of three.

Y at a moment of future time. When we say that X is Y (in the temporal sense of 'is'), we are asserting X to be Y at a moment of present time.

Thus our first statement about M—that it is present, will be past, and has been future—means that M is present at a moment of present time, past at some moment of future time, and future at some moment of past time. But every moment, like every event, is both past, present, and future. And so a similar difficulty arises. If M is present, there is no moment of past time at which it is past. But the moments of future time, in which it is past, are equally moments of past time, in which it cannot be past. Again, that M is future and will be present and past means that M is future at a moment of present time, and present and past at different moments of future time. In that case it cannot be present or past at any moments of past time. In that case it cannot be present or past at any moments of past time. But all the moments of future time, in which M will be present or past, are equally moments of past time.

And thus again we get a contradiction, since the moments at which M has any one of the three determinations of the A series are also moments at which it cannot have that determination. If we try to avoid this by saying of these moments what had been previously said of M itself—that some moment, for example, is future, and will be present and past—then 'is' and 'will be' have the same meaning as before. Our statement, then, means that the moment in question is future at a present moment, and will be present and past at different moments of future time. This, of course, is the same difficulty over again. And so on infinitely.

Such an infinity is vicious. The attribution of the characteristics past, present, and future to the terms of any series leads to a contradiction, unless it is specified that they have them successively. This means, as we have seen, that they have them in relation to terms specified as past, present, and future. These again, to avoid a like contradiction, must in turn be specified as past, present, and future. And, since this continues infinitely, the first set of terms never escapes from contradiction at all.[6]

The contradiction, it will be seen, would arise in the same way supposing that pastness, presentness, and futurity were original qualities, and not, as we have decided that they are, relations. For it would still be the case that they were characteristics which were incompatible with one another, and that whichever had one of them would also have the other. And it is from this that the contradiction arises.

---

[6] It may be worth while to point out that the vicious infinite has not arisen from the impossibility of *defining* past, present, and future, without using the terms in their own definitions. On the contrary, we have admitted these terms to be indefinable. It arises from the fact that the nature of the terms involves a contradiction, and that the attempt to remove the contradiction involves the employment of the terms, and the generation of a similar contradiction.

The reality of the A series, then, leads to a contradiction, and must be rejected. And, since we have seen that change and time require the A series, the reality of change and time must be rejected. And so must the reality of the B series, since that requires time. Nothing is really present, past, or future. Nothing is really earlier or later than anything else or temporally simultaneous with it. Nothing really changes. And nothing is really in time. Whenever we perceive anything in time—which is the only way in which, in our present experience, we do perceive things—we are perceiving it more or less as it really is not.[7]

[7] Even on the hypothesis that judgements are real it would be necessary to regard ourselves as perceiving things in time, and so perceiving them erroneously.

II

CHANGES IN EVENTS AND CHANGES IN THINGS

ARTHUR N. PRIOR

The basic question to which I wish to address myself in this lecture is simply the old one, does time really flow or pass? The problem, of course, is that genuine flowing or passage is something which occurs *in* time, and *takes* time to occur. If time itself flows or passes, must there not be some 'super-time' in which it does so? Again, whatever flows or passes does so at some *rate*, but a rate of flow is just the amount of movement in a given *time*, so how could there be a rate of flow of time itself? And if time does not flow at any rate, how can it flow at all?

A natural first move towards extricating ourselves from these perplexities is to admit that talk of the flow or passage of time is just a metaphor. Time may be, as Isaac Watts says, *like* an ever-rolling stream, but it isn't really and literally an ever-rolling stream. But *how* is it like an ever-rolling stream? What is the literal truth behind this metaphor? The answer to this is not, at first sight, difficult. Generally when we make such remarks as 'Time does fly, doesn't it?—why, it's already the 16th', we mean that some date or moment which we have been looking forward to as future has ceased to be future and is now present and on its way into the past. Or more fundamentally, perhaps, some future *event* to which we have been looking forward with hope or dread is now at last occurring, and soon will have occurred, and will have occurred a longer and longer time *ago*. We might say, for example, 'Time does fly—I'm already 47'—that is, my birth is already that much past, 'and soon I shall be 48', i.e. it will be more past still. Suppose we speak about something 'becoming more past' not only when it moves from the comparatively near past to the comparatively distant past, but also when it moves from the present to the past, from the future to the present, and from the comparatively distant future to the comparatively near future. Then whatever is happening, has happened, or will happen is all the time 'becoming more past' in this extended sense; and just this is what we mean by the flow or passage of time. And if we want to give the *rate* of this flow or passage, it

Arthur N. Prior, *Papers on Time and Tense* (Oxford: Clarendon Press, 1968), ch. 1. © Oxford University Press 1968. Reprinted by permission of Oxford University Press.

is surely very simple—it takes one exactly a year to get a year older, i.e. events become more past at the rate of a year per year, an hour per hour, a second per second.

Does this remove the difficulty? It is far from obvious that it does. It's not just that an hour per hour is a queer sort of rate—*this* queerness, I think, has been exaggerated, and I shall say more about it in a minute—but the whole idea of events changing is at first sight a little strange, even if we abandon the admittedly figurative description of this change as a *movement*. By and large, to judge by the way that we ordinarily talk, it's *things* that change, and events don't change but *happen*. Chairs, tables, horses, people change— chairs get worn out and then mended, tables get dirty and then clean again, horses get tired and then refreshed, people learn things and forget them, or are happy and then miserable, active and then sleepy, and so on, and all these are changes, and chairs, tables, horses, and people are all what I mean by things as opposed to events. An accident, a coronation, a death, a prize- giving, are examples of what we'd call events, and it does seem unnatural to describe these as changing—what these do, one is inclined to say, is not to change but to happen or occur.

One of the things that make us inclined to deny that events undergo changes is that events *are* changes—to say that such and such an event has occurred is generally to say that some thing has, or some things have, changed in some way. To say, for instance, that the retirement of Sir Anthony Eden occurred in such and such a year is just to say that Sir Anthony then retired and so suffered the change or changes that retirement consists in—he had been Prime Minister, and then was not Prime Minister. Sir Anthony's retirement is or was a change concerning Sir Anthony; to say that it itself changes or has changed sounds queer because it sounds queer to talk of a change changing.

This queerness, however, is superficial. When we reflect further we realize that changes do change, especially if they go on for any length of time. (In this case we generally, though not always, call the change a *process* rather than an event, and there are other important differences between events and processes besides the length of time they take, but these differences are not relevant to the present discussion, so I shall ignore them and discuss changes generally, events and processes alike.) Changes do change—a movement, for example, may be slow at first and then rapid, a prize-giving or a lecture may be at first dull and afterwards interesting, or vice versa, and so on. It would hardly be too much to say that modern science began when people became accustomed to the idea of changes changing, e.g. to the idea of acceleration as opposed to simple motion. I've no doubt the ordinary measure of acceler- ation, so many feet per second per second, sounded queer when it was first used, and I think it still sounds queer to most students when they first encounter it. Ordinary speech is still resistant to it, and indeed to the expression of anything in the nature of a comparison of a comparison. We are taught at school that 'more older', for example, is bad English, but why shouldn't I say that I am more older than my son than he is than my daughter? And if we have learned to talk of an acceleration of a foot per second per second without imagining that the second 'second' must somehow be a different kind of 'second' from the first one—without imagining that if motion takes place in ordinary time, acceleration must take place in some super-time—can we not accustom ourselves equally to a change of 'a second per second' without any such imagining?

Changes do change, then, but this does not leave everything quite simple and solved. For there's still something odd about the change that we describe figuratively as the flow or passage of time—the change from an event's being future to its being present, and from its being present to its being more and more past. For the other changes in events which I have mentioned are ones which go on in the event *while it is occurring*; for example, if a lecture gets duller or a movement faster then this is something it does *as it goes on*; but the change from past to still further past isn't one that occurs while the event is occurring, for all the time that an event is occurring it isn't past but present, in fact the presentness of an event just *is* its happening, its occurring, as opposed to its merely having happened or being merely about to happen. We might put it this way: the things that change are *existing* things, and it's while they exist that they change, e.g. it's existing men, not non-existent men, that get tired and then pick up again; Julius Caesar, for example, isn't now getting tired and picking up again, unless the doctrine of immortality is true and he exists now as much as he ever did. And such changes as the change in the rate of movement are similarly changes that go on in events or processes while they exist, that is, while they exist in the only sense in which events and processes do exist, namely while they are occurring. But getting more and more past seems to be something an event does when it *doesn't* exist, and this seems very queer indeed.

We may retrace our steps to this point by looking at some of the literature of our subject. Professor C. D. Broad, in the second volume of his *Examination of McTaggart's Philosophy*, says that the ordinary view that an event, say the death of Queen Anne, is in the indefinitely distant future and then less and less future and then present and then goes into the more and more distant past—this ordinary story, Broad says, cannot possibly be true because it takes the death of Queen Anne to be at once a mere momentary thing and something with an indefinitely long history. We can make a first answer to this by distinguishing between the history that an event *has*, and the bit of history that it *is*. The bit of history that Queen Anne's death is, or was, is a very very short bit, but that doesn't prevent the history that it has from being indefinitely long. Queen Anne's death is part of the history of Queen Anne, and a very short part of it; what is long is not this part of the history of Queen Anne, but rather the history of this part of her history—the history of this part of her history is that first it was future, then it was present, and so on, and this can be a long history even if the bit of history that it is the history *of* is very short. There is not, therefore, the flat contradiction that Broad suggests here. There is, however, the difficulty that we generally think of the history of a thing as the sum of what it does and what happens to it *while it is there*—when it ceases to be, its history has ended—and this does make it seem odd that there should be an indefinitely long history of something which itself occupies a time which is indefinitely short.

But if there is a genuine puzzle here, it concerns what is actually going on also. For whatever goes on for any length of time—and that means: whatever goes on—will have future and past phases as well as the immediately present one; its going on is in fact a continual passage of one phase after another from being future through being present to being past. Augustine's reflections, in the eleventh chapter of his *Confessions*, on the notion of a 'long time', are relevant here. Just when, he asks, is a long time long? Is it long when it is present, or when it is past or future? We need not, I think, attach much importance to the fact that Augustine concentrates on so abstract a thing as a 'time' or an interval; his problems can be quite easily restated in terms of *what goes on* over the interval; in fact he himself slips into this, and talks about his childhood, a future sunrise, and so on. When, we may ask, does a process go on for a long time—while it is going on, or when it lies ahead of us, or is all over?

Augustine is at first driven to the view that it is when it is present that a time is long, for only what *is* can be long or short (paragraph 18). We can give the same answer with processes—it is when they are going on that they go on for a long time. But then, as Augustine points out, there are these phases. A hundred years is a long time, but it's not really present all at once, and even if we try to boil down the present to an hour, 'that one hour passes away in flying particles'. 'The present hath no space' (20). Augustine had apparently not heard of the 'specious present', but even if he had it would not have helped him much—most of the happenings we are interested in take longer than that. He tries out the hypotheses that the past and the future, and past and future events, in some sense after all 'are'—that there is some 'secret place' where they exist all the time, and from which they come and to which they go. If there is no such place, then where do those who foresee the future and recall the past discern these things? 'For that which is not, cannot be seen' (22).

Well, Augustine says, he doesn't know anything about that, but one thing that he does know is that wherever 'time past and to come' may 'be', 'they are not there as future, or past, but present. For if there also they be future, they are not yet there; if there also they be past, they are no longer there. Wheresoever then is whatsoever is, it is only as present' (23). Of course there are present 'traces' or images of past things in our memories, and present signs and intentions on the basis of which we make our future forecasts (23, 24), and sometimes Augustine seems satisfied with this—past, present, and future, he says, 'do exist in some sort, in the soul, but otherwhere do I not see them' (26). But sometimes he seems far from content with this—*that which* we remember and anticipate, he says, is different from these signs, and is *not* present (23, 24)—and, one must surely add, is *not* 'in the soul'.

It is time now to be constructive, and as a preparation for this I shall indulge in what may seem a digression, on the subject of Grammar. English philosophers who visit the United States are always asked sooner or later whether they are 'analysts'. I'm not at all sure what the answer is in my own case, but there's another word that Professor Passmore once invented to describe some English philosophers who are often called 'analysts', namely the word 'grammaticist', and that's something I wouldn't at all mind calling myself. I don't deny that there are genuine metaphysical problems, but I think you have to talk about grammar at least a little bit in order to solve most of them. And in particular, I would want to maintain that most of the present group of problems about time and change, though not quite all of them, arise from the fact that many expressions which look like nouns, i.e. names of objects, are not really nouns at all but concealed verbs, and many expressions which look like verbs are not really verbs but concealed conjunctions and adverbs. That is a slight over-simplification, but before we can get it stated more accurately we must look more closely at verbs, conjunctions, and adverbs.

I shall assume that we are sufficiently clear for our present purposes as to what a noun or name is, and what a sentence is; and given these notions, we can define a verb or verb-phrase as an expression that constructs a sentence out of a name or names. For instance, if you tack the verb 'died' on the name 'Queen Anne' you get the sentence 'Queen Anne died', and if you tack the phrase 'is an undertaker' on the name 'James Bowels' you get the sentence 'James Bowels is an undertaker', so that this is a verb-phrase. I say 'out of a name *or names*' because some verbs have to have an object as well as a subject. Thus if you put the verb 'loves' between the names 'Richard' and 'Joan' you get the sentence 'Richard loves Joan'; this verb constructs this sentence out of these two names; and the phrase 'is taller than' would function similarly. Logicians call verbs and verb-phrases 'predicates'; 'died' and 'is an undertaker' would be 'one-place' predicates, and 'loves' and 'is taller than' are 'two-place' predicates. There are also expressions which construct sentences, not out of names, but out of other sentences. If an expression constructs a sentence out of two or more other sentences it is a conjunction, or a phrase equivalent to a conjunction. For example 'Either—or—' functions in this way in 'Either it will rain or it will snow'. If the expression constructs a sentence out of one other sentence it is an adverb or adverbial phrase, like 'not' or 'It is not the case that', or 'allegedly' or 'It is alleged that', or 'possibly' or 'It is possible that'. Thus by attaching these expressions to 'It is raining' we obtain the sentences

It is not raining;
It is not the case that it is raining;
It is allegedly raining;
It is alleged that it is raining;
It is possibly raining;
It is possible that it is raining.

One very important difference between conjunctions and adverbs, on the one hand, and verbs, on the other, is that because the former construct sentences out of sentences, i.e. the same sort of thing as they end up with, they can be applied again and again to build up more and more complicated sentences, like 'It is allegedly possible that he will not come', which could be spread out as

It is said that (it is possible that (it is not the case that (he will come))).

You can also use the same adverb twice and obtain such things as double negation, alleged allegations, and so on. Verbs, because they do not end up with the same sort of expression as what they start with, cannot be piled up in this way. Having constructed 'Queen Anne died' by the verb 'died' out of the name 'Queen Anne', you cannot do it again—'Queen Anne died died' is not a sentence.

Turning now to our main subject, I want to suggest that putting a verb into the past or future tense is exactly the same sort of thing as adding an adverb to the sentence. 'I *was* having my breakfast' is related to 'I am having my breakfast' in exactly the same way as 'I am *allegedly* having my breakfast' is related to it, and it is only an historical accident that we generally form the past tense by modifying the present tense, e.g. by changing 'am' to 'was', rather than by tacking on an adverb. In a rationalized language with uniform constructions for similar functions we could form the past tense by prefixing to a given sentence the phrase 'It was the case that', or 'It has been the case that' (depending on what sort of past we meant), and the future tense by prefixing 'It will be the case that'. For example, instead of 'I will be eating my breakfast' we could say

It will be the case that I am eating my breakfast,

and instead of 'I was eating my breakfast' we could say

It was the case that I am eating my breakfast.

The nearest we get to the latter in ordinary English is 'It was the case that I *was* eating my breakfast', but this is one of those anomalies like emphatic double negation. The construction I am sketching embodies the truth behind Augustine's suggestion of the 'secret place' where past and future times 'are', and his insistence that wherever they are, they are not there as past or future but as present. The past is not the present but it *is* the past present, and the future is not the present but it *is* the future present.

There is also, of course, the past future and the future past. For these adverbial phrases, like other adverbial phrases, can be applied repeatedly— the sentences to which they are attached do not have to be simple ones; it is enough that they be sentences, and they can be sentences which already have tense-adverbs, as we might call them, within them. Hence we can have such a construction as

It will be the case that (it has been the case that (I am taking off my coat)),

or in plain English, 'I will have taken off my coat'. We can similarly apply repeatedly such *specific* tense-adverbs as 'It was the case forty-eight years ago that'. For example, we could have

It will be the case seven months hence that (it was the case forty-eight years ago that (I am being born)),

that is, it will be my forty-eighth birthday in seven months' time.

To say that a change has occurred is to say at least this much: that something which was the case formerly is not the case now. That is, it is at least to say that for some sentence *p* we have

It was the case that *p*, and it is not the case that *p*.

This sentence *p* can be as complicated as you like, and can itself contain tense-adverbs, so that one example of our formula would be

It was the case five months ago that (it was the case only forty-seven years ago that (I am being born)), and it is not now the case that (it was the case only forty-seven years ago that (I am being born)), that is, I am not as young as I used to be. This last change, of course, is a case of precisely that recession of events into the past that we are really talking about when we say that time flows or passes, and the piling of time-references on top of one another, with no suggestion that the time-words must be used in a different sense at each level, simply reflects the fact that tense-adverbs *are* adverbs, not verbs.

An important point to notice now is that while *I* have been talking about words—for example, about verbs and adverbs—for quite a long time, the sentences that I have been using as examples have *not* been about words but about real things. When a sentence is formed out of another sentence or other sentences by means of an adverb or conjunction, it is not *about* those other sentences, but about whatever they are themselves about. For example, the compound sentence 'Either I will wear my cap or I will wear my beret' is not about the sentences 'I will wear my cap' and 'I will wear my beret'; like them, it is about me and my headgear, though the information it conveys about these is a little less definite than what either of them would convey separately. Similarly, the sentence 'It will be the case that I am having my tooth out' is not about the sentence 'I am having my tooth out'; it is about me. A genuine sentence about the sentence 'I am having my tooth out' would be one stating that it contained six words and nineteen letters, but 'It will be the case that I am having my tooth out', i.e. 'I will be having my tooth out', is quite obviously not a sentence of this sort at all.

Nor is it about some abstract entity named by the clause 'that I am having my tooth out'. It is about me and my tooth, and about nothing else whatever. The fact is that it is difficult for the human mind to get beyond the simple subject–predicate or noun–verb structure, and when a sentence or thought hasn't that structure but a more complex one we try in various ways to force it into the subject–predicate pattern. We thus invent new modes of speech in which the subordinate sentences are replaced by noun-phrases and the conjunctions or adverbs by verbs or verb-phrases. For example, instead of saying

(1) *If* you have oranges in your larder you have been to the greengrocer's,

we may say

(2) Your having oranges in your larder *implies* your having been to the greengrocer's,

which looks as if it has the same form as 'Richard loves Joan' except that 'Your having oranges in your larder' and 'Your having been to the grocer' seem to name more abstract objects than Richard and Joan, and implying seems a more abstract activity than loving. We can rid ourselves of this suggestion if we reflect that (2) is nothing more than a paraphrase of (1). Similarly,

(3) It is now six years since it was the case that I am falling out of a punt,

could be rewritten as

(4) My falling out of a punt has receded six years into the past.

This suggests that something called an event, my falling out of a punt, has gone through a performance called receding into the past, and moreover has been going through this performance even after it has ceased to exist, i.e. after it has stopped happening. But of course (4) is just a paraphrase of (3), and like (3) is not about any objects except me and that punt—there is no real reason to believe in the existence either now or six years ago of a further object called 'my falling out of a punt'.

What I am suggesting is that what looks like talk about events is really at bottom talk about things, and that what looks like talk about changes in events is really just slightly more complicated talk about changes in things. This applies *both* to the changes that we say occur in events when they are going on, like the change in speed of a movement ('movement' is a *façon de parler*; there is just the moving car, which moves more quickly than it did), *and* the changes that we say occur in events when they are not going on any longer, or not yet, e.g. my birth's receding into the past ('birth' is a *façon de parler*—there's just me being born, and then getting older).

It's not all quite as simple as this, however. This story works very well for me and my birth and my fall out of the punt, but what about Queen Anne? Does Queen Anne's death getting more past mean that *Queen Anne* has changed from having died 250 years ago to having died 251 years ago, or whatever the period is?—that *she* is still 'getting older', though in a slightly extended sense? The trouble with this, of course, is just that Queen Anne doesn't exist now any more than her death does. There are at least two different ways in which we might deal with this one. We might, in the first place, say that our statement really is about Queen Anne (despite the fact that she 'is no more'), and really is, or at least entails, a statement of the form

It was the case that *p*, and is not now the case that *p*,

namely

It was the case that it was the case only 250 years ago that Queen Anne is dying, and is not now the case that it was the case only 250 years ago that Queen Anne is dying, but we may add that this statement does not record a 'change' in any natural sense of that word, and certainly not a change in Queen Anne. A genuine record of change, we could say, must not only be of the form above indicated but must meet certain further conditions which we might specify in various ways. And we could say that although what is here recorded *isn't* a change in the proper sense, it is *like* a change in fitting the above formula. The flow of time, we would then say, is merely metaphorical, not only because what is meant by it isn't a genuine movement, but further because what is meant by it isn't a genuine change; but the force of the metaphor can still be explained—we use the metaphor because what we call the flow of time does fit the above formula. On this view it might be that not only the recession of Queen Anne's death but my own growing older will not count as a change in the strict sense, though growing older is normally *accompanied* by genuine changes, and the phrase is commonly extended to cover these—increasing wisdom, bald patches, and so on.

But can a statement really be *about* Queen Anne after she has ceased to be? I do not wish to dogmatize about this, but an alternative solution is worth mentioning. We might paraphrase 'Queen Anne has died' as 'Once there was a person named "Anne", who reigned over England, etc., but there is not now any such person'. This solution exploits a distinction which we may describe as one between *general facts* and *individual facts*. That someone has stolen my pencil is a general fact; that John Jones has stolen my pencil, if it is a fact at all, is an individual fact. It has often been said—for example, it was said by the Stoic logicians—that there are no general facts without there being the corresponding individual facts. It cannot, for example, be the case that 'someone' has stolen my pencil, unless it is the case that some specific individual—if not John Jones, then somebody else—has stolen it. And in cases of this sort the principle is very plausible, indeed it is obviously true. I have read that some of the schoolmen described the subject of sentences like 'someone has stolen my pencil' as an *individuum vagum*, but of course this is a makeshift—forcing things into a pattern again. There are no 'vague individuals', and if a pencil has been stolen at all it has been stolen not by a vague individual but by some quite definite one, or else by a number of such. There are vague statements, however, and vague thoughts, and the existence of such statements and thoughts is as much a fact about the real world as any other; and when we describe the making of such statements and the enter- taining of such thoughts, we do encounter at least partly general facts to which no wholly individual facts correspond. If I allege or believe that someone has stolen my pencil, there may be *no* specific individual with respect to whom I allege or believe that *he* stole my pencil. There is *alleged or believed to be* an individual who stole it, but there is *no individual who is alleged or believed* to have stolen it (not even a vague one). So while it is a fact that I allege or believe that someone stole it, there is no fact of the form 'I allege (or believe) that X stole it'. The one fact that there is is no doubt an individual fact in so far as it concerns me, but is irreducibly general as far as the thief is concerned. (There may indeed be *no* thief—I am perhaps mistaken about the whole thing—but this is another question; our present point is that there may be no one who is even said or thought to be a thief, though it is said or thought *that there is* a thief.)

Returning now to Queen Anne, what I am suggesting is that the sort of thing that we unquestionably do have with 'It is said that' and 'It is thought that', we also have with 'It will be the case that' and 'It was the case that'. It *was the case that someone* was called 'Anne', reigned over England, etc., even though *there is not now anyone* of whom it was the case that *she* was called 'Anne', reigned over England, etc. What we must be careful about here is simply getting our prefixes in the right order. Just as

(1) I think that (for some specific X (X stole my pencil))

does not imply

(2) For some specific X (I think that (X stole my pencil)),

so

(3) It was the case that (for some specific X (X is called 'Anne', reigns over England, etc.))

does not imply

(4) For some specific X (it was the case that (X is called 'Anne', reigns over England, etc.)).

On this view, the fact that Queen Anne has been dead for some years is not, in the strict sense of 'about', a fact about Queen Anne; it is not a fact about anyone or anything—it is a *general* fact. Or if it is about anything, what it is about is not Queen Anne—it is about the earth, maybe, which has rolled around the sun so many times since there was a person who was called 'Anne', reigned over England, etc. (It would then be a *partly* general fact—individual in so far as it concerns the earth, but irreducibly general as far as the dead queen is concerned. But if there are—as there undoubtedly are—irreducibly partly general facts, could there not be irreducibly wholly general ones?) Note, too, that the fact that this fact is not about Queen Anne cannot itself be a fact about Queen Anne—its statement needs rephrasing in some such way as 'There is no person who was called "Anne", etc., and about whom it is a fact that, etc.'

On this view, the recession of Queen Anne's death into the further past is quite decidedly not a change in Queen Anne, not because we are using 'change' in so tight a sense that it is not a change at all, but because Queen Anne doesn't herself enter into this recession, or indeed, now, into any fact whatever. But the recession *is* still a change or quasi-change in the sense that it fits the formula 'It was the case that *p*, but is not now the case that *p*'—this formula continues to express what is common to the flow of a literal river on the one hand (where it was the case that such and such drops were at a certain place, and this is the case no longer) and the flow of time on the other.

III

THE UNREALITY OF TENSE

D. H. MELLOR

Change is clearly of time's essence, and many have thought it the downfall of the tenseless view of time—that only a tensed view of time can account for it. In fact the opposite is true. The reality of tense is disproved by a contradiction inherent in the idea that time flows, i.e. that things, events, and facts really change when their tenses (i.e. locations in McTaggart's A series) change from future to present to past.

Change, obviously (if vaguely), is having a property at one time and not at another. More specifically, it is something having incompatible properties—such as being at different temperatures or in different places—at different dates (i.e. locations in McTaggart's B series). Thus cooling is a change of temperature: something's being first hot and then cold. Movement, likewise, is a change of place: something's being first in one place and then in another. Similarly, there are changes in the sizes, shapes, colours, and other properties of things. In each case something has one of several mutually incompatible properties at one B series time and another one later.

This tenseless idea of change is basically right as well as being obvious. But there are objections to it, of which two especially have long preserved a tensed alternative. The first is that it does not really distinguish change through time from change across space. Properties can after all vary from place to place as well as from time to time. A poker, for example, may be simultaneously hot at one end and cold at the other: why is that not change, as much as the whole poker being hot one day and cold the next? We could of course define change to be variation in time as opposed to variation in space—but only given some other way of distinguishing time from space. If time is marked off only as the dimension of change, we should be arguing in an indecently small circle. But it is not obvious how else to distinguish time from the dimensions of space.

Consider a clock's second hand passing successively the figures '1' and '2'. The latter event is both later than the former and to the right of it. We

D. H. Mellor, *Real Time* (Cambridge: Cambridge University Press, 1981), ch. 6, revised by the author for this volume. Reprinted by permission of Cambridge University Press.

see this as change (namely, movement), which is how we distinguish the temporal and spatial relations of the two events. Specifically, it is how we tell a thin hand moving across the clock face from a fat one spread statically across it in two spatial dimensions. In short, we perceive the temporal relation in this case by perceiving change. Similarly for changes of place, temperature, colour, and everything else. To see that one event is later than another is to see something change. How else could time be perceived or understood, except as the dimension of change? But change cannot then be defined as variation in time. And the objection to tenseless time is that it has no other way of defining change, and so in particular no way of distinguishing change from spatial variation.

This objection is reinforced by the other one, namely that the tenseless view of time reduces change to changeless facts. If a poker is hot one day and cold the next, then those always were and always will be its temperatures at those two dates. B series facts of this kind do not change with time: that after all being the mark of the B as opposed to the A series. And as for time, so for space. There is no spatial analogue of change in a poker's being simultaneously hot at one end and cold at the other. The hot and cold ends of a poker are not a case of 'spatial change', because they coexist: what we might call the 'spatially tenseless' world contains them both. The ends of the poker are simply differently located in tenseless space, i.e. in the space of maps, which identify places without reference to the spatial present (*here*). Likewise, the hot poker and the cold coexist in a temporally tenseless world. It contains them both, only located in different parts of tenseless time. And if, as everyone agrees, coexistence prevents change in the spatial case, how can it be compatible with change in time?

The advocate of tensed time has a ready answer to these questions. Change, he says, is basically the changing tense (A series location) of things and events moving from future to past. It is peculiar to time because the A series has no real spatial analogue. In other words, the spatially tenseless 'token-reflexive' truth-conditions of 'spatially tensed' thoughts and sentences (e.g. that for any place x what makes anyone 'token'—i.e. think, say, or write—'x is here' truly is that they do so at or within x) are all there is to spatial tense. There are not also 'spatially tensed' facts (like x being here) which are facts only in some places and not in others. But on the tensed view of time the temporally tenseless token-reflexive truth-conditions (e.g. that for any B series location t what makes any token of 't is now' true is its occurring at or within t) are *not* all there is to temporal tense. In the temporal case, over and above all such tenseless facts, there are tensed facts (like t being now) which are facts only at some B series times and not at others. In particular, there is a present moment (*now*) which is forever changing its date (t). The reality of the clock hand's movement consists ultimately in the events of its passing the figures '1' and '2' becoming successively present and then past; and similarly for all other changes.

On the tensed view, then, change is primarily the successive temporal presence of earlier and later things and events. This defines the tenseless temporal relation *earlier* (and hence *later*): one event is earlier than another if and only if its ever-changing tenses make it present first. And if there are no real spatial tenses, this definition has no spatial analogue. The tenseless spatial relations of events and things are *sui generis*—and that is the difference between space and time. In time, but not in space, the tenseless *B* series, and hence the idea of change I started with, is supposed to be derived from the *A* series. Change is still defined as variation through time; but by defining time first as the dimension of changing tense, the tensed view prevents this definition of change begging the question against its spatial analogues.

This tensed view of change may be supported by further doctrines about what else turns on the difference between present and future. There is, for example, the view that existence turns on it, i.e. that coming to be present is coming to exist. This provides a still more profound basis for distinguishing time from space, as the dimension in which things come successively into existence. However, it makes no odds what else turns on tense unless the idea of changing tenses actually does account for change; and, despite first appearances to the contrary, the fact is that it doesn't.

To start with, one might accuse the tensed view itself of begging the question against spatial change by denying the reality of spatial tenses. Spatial tenses like the hereness of x do after all vary across space just as temporal tenses like the nowness of t vary through time. If change is different events being temporally present at different times, why is it not different things being spatially present—here—at different places? To such accusations upholders of tense reply that we have a direct intuition of temporal presence which is lacking in the spatial case. We see things laid out tenselessly in space, they say, whereas we do not see things laid out tenselessly in time.

But this is not true. When we see the tenseless order of two events (e.g. a clock hand passing '1' earlier than it passes '2'), we don't thereby see their tenses. Astronomers who observe that one celestial event is earlier than another don't do so by seeing that it's more past. Celestial events don't look as past as they are (if they did, cosmology would be a great deal easier than it is): indeed they don't *look* past at all. Tense is not a perceptible property of the things and events we perceive. One cannot, for example, refute a fortune-teller who claims to see the future by pointing to the visible pastness of what he sees in his crystal ball, since it will look just the same wherever it is in the A series. The fact is that we only know the A series whereabouts of what we see by knowing how much earlier it is than our seeing them—an experience which as we have it we admittedly know we're having *now*, but only for the same trivial token-reflexive reason that we know we're having it *here*.

In short, we see things varying through time just as tenselessly as we see them varying across space. But although the tensed view of change could easily be convicted in this and other ways of distinguishing time from space no better than the tenseless view, I will not press that charge here. To prefer debatable and relatively trifling charges is pointless when a capital offence can be proved against the same party. The capital charge I want to press here is self-contradiction, an offence of which tensed views of time and space are equally guilty. But as no one will defend the tensed view of space, I need only give the prosecution's case against the tensed view of time.

The proof of contradiction in the tensed account of change is not new. It was given by J. E. McTaggart in 1908 and has been much debated since. To me it seems beyond all reasonable doubt, but, since it is still disputed, I fear I must present it yet again. Two factors, however, encourage me to hope for more success than McTaggart had. One is that he attacked time itself. He thought both that time needs change and that change needs changing tense, and so thought to convict time along with tense. However, the obvious reality of time defeated him, and unfortunately drew suspicion on his whole argument. Tense in short has been wrongly acquitted to save the innocent time. But we need not acquit the guilty in this case in order to save the innocent. What is wrong with McTaggart's attack on time is not his attack on tense but his contention that disposing of tense disposes of change. Change can in fact be explained and distinguished from spatial variation without any appeal to tense (see my *Real Time*,[1] ch. 7). And, given that, the reality of changing tense can safely be denied without imperilling the reality of change and hence of time itself. Once this is realized, McTaggart's proof will, I hope, meet much less resistance.

The other factor encouraging me is the now standard account of the tenseless token-reflexive truth-conditions of tensed thoughts and sentences. Although this factor is not new, it was not there in McTaggart's time, and it should make his proof more persuasive. For, on the one hand, it should make the validity of his proof more obvious and, on the other, its conclusions more palatable. Tense, it turns out, is not being banished altogether, merely replaced where it belongs: namely, in our heads, as a way of thinking which

---

[1] Cambridge: Cambridge University Press, 1981.

we need in order to be capable of timely action. But that is another story (see *Real Time*, ch. 5).

McTaggart's proof is very simple. Many *A* series locations are incompatible with each other. An event which is *yesterday*, for example, cannot also be *tomorrow*. Past, present, and future tenses are mutually incompatible properties of things and events. But because they are forever changing, everything has to have them all. Everything occupies every *A* series location, from the remotest future, through the present, to the remotest past. But nothing can really have incompatible properties, so nothing in reality has tenses. The *A* series is a myth.

The defence has an immediate and obvious riposte to this attack, and its rebuttal is unfortunately much less obvious; which is why McTaggart's proof has rarely carried the conviction it deserves. The riposte is that nothing has incompatible tenses at the same time. Nothing is present *when* it is past, or future when it is present. Things and events only have these properties successively: first they are future, then present, then past. And nothing prevents things from having incompatible properties at different times. On the contrary, that is how change is defined: the successive possession of incompatible properties. All McTaggart has shown is that changing tense fits that definition, as it should.

To rebut this riposte, McTaggart asks when, in tensed terms, things and events have their various tenses; and here it will help to use some symbols. Let P, N, and F be respectively the properties of being past, present (i.e. now), and future, and let *e* be any event. Then *e* being past, present, and future I write respectively as 'P*e*', 'N*e*', and 'F*e*'. Complex tenses I represent by repeated ascription of P, N, and F: thus 'PF*e*' means *e was* future, 'FPN*e*' means *e will have been* present, and so on. '~', '&', and '⊢' as usual mean respectively 'not', 'and', and 'entails'.

Then McTaggart's basic argument is that, on the one hand, the three properties P, N, and F are mutually incompatible, so that for any event *e*

(1) P*e* ⊢ ~ N*e*; N*e* ⊢ ~ F*e*; F*e* ⊢ ~ P*e*; etc.

On the other hand, the inexorable change of tense means that every event has all three *A* series locations, i.e.

(2) P*e* & N*e* & F*e*.

But (1) and (2) cannot both be true; since if (1) is true, two of the statements in (2) must be false, so (2) as a whole must be false. But our concept of tense commits us to both (1) and (2); so it leads us inevitably into contradiction and thus cannot apply to reality. Reality therefore must be tenseless: there are no tensed facts.

To this the riposte is that e has no more than one of these incompatible properties at once, so there is no contradiction after all. Suppose, for example, that e is actually present, i.e. Ne. Then e is neither past nor future, i.e. both 'Pe' and 'Fe' are false, as (1) requires. The truth rather is that e *will be* past and *was* future, i.e. not (2) but

(3) FPe & Ne & PFe,

which is quite compatible with (1).

So it is. But, as McTaggart remarks, there are more complex tenses than those in (3), and not all combinations of them are so innocuous. Specifically, there are also PP and PN, FF and FN, and NP, NN, and NF. And just as every event has all A series locations if it has any of them, so it also has all these other complex tenses. For example, whatever has any simple tense obviously also has it *now*, i.e.

Pe ⊢ NPe; Ne ⊢ NNe; Fe ⊢ NFe.

Obviously, also, whatever is past *was* present and *was* future, and whatever is future *will be* present and *will be* past, so that

Pe ⊢ PNe; Pe ⊢ PFe; Fe ⊢ FNe; Fe ⊢ FPe.

Moreover, whatever is sufficiently past also *was* past, e.g. what happened two days ago was already past yesterday; and sufficiently future events likewise also *will be* future: which gives us PP and FF as well as P and F.

In place then of the original three simple tenses, we have the nine compound tenses PP, PN, PF; NP, NN, NF; FP, FN, FF. But McTaggart's argument applies just as well to them. Because of the way tense incessantly changes, every event that has any of these nine tenses has to have them all; but they are not all mutually compatible. For example, FF and PP are incompatible, since what will be future cannot also have been past. And NP, NN, and NF are even more clearly incompatible, because they are equivalent to the simple P, N, and F.

The riposte will again be made that events do not have these incompatible tenses all at once. But again, saying in tensed terms just when they do have them only generates another set of properties, including mutually incompatible ones like PPP, NNN, and FFF, all of which every event has to have. There is, in other words, an endless regress of ripostes and rebuttals, a regress that is vicious because at no stage in it can all the supposed tensed facts be consistently stated.

THE DEFENCE OF McTAGGART

That, basically, is how McTaggart put his case. His critics have reacted by denying the viciousness of his regress. At every stage, they say, the appearance of contradiction is removed by distinguishing the different times at which events have different tenses. They ignore the fact that the tensed means they use to distinguish these times are also subject to the contradiction they are trying to remove. However, the debate by now is too well worn to be settled by mere repetition of McTaggart's proof, sound though it is. To change the metaphor, too many people have managed to inoculate themselves against it. If it is to wipe out belief in real tenses, as it should, a more immediately virulent strain of it is needed, a strain that I believe is best nurtured on the token-reflexive facts that make tensed sentences true or false.

Before developing the new strain, however, it is worth neutralizing some antidotes that have been proposed to McTaggart's original proof. First, I should perhaps remark that although I have dealt only with the unqualified past, present, and future, the proof applies also to more precise *A* series locations. *Yesterday* and *three days ago*, for example, are likewise incompatible properties of things and events of less than a day's duration, both of which they must all, nevertheless, possess. But there is no point in complicating the discussion by bringing all these other tenses into it explicitly. If the argument works for P, N, and F, it will work for all tenses; and if not, for none.

Secondly, I have followed McTaggart in ascribing these problematic *A* series properties to events. Tense logicians mostly prefer to treat 'P', 'N', and 'F' as 'operators' (analogous to 'It is not the case that' or 'It may be the case that') prefixed to present-tense core sentences or propositions. This is tantamount to regarding P, N, and F as properties, not of events, but of tensed facts. Where, for example, McTaggart and I start with a thunderstorm, tense logic starts with the sentence or proposition saying that a thunderstorm is happening now. Where we say the thunderstorm is two days past, they say the fact of its happening now is two days past, i.e. the present-tense sentence or proposition saying that it is happening now was true two days ago. In the symbolism above, this amounts to replacing '*e*' throughout by '*Ne*'. But it makes no odds to the argument, as readers may verify for themselves: facts are no better at being at once both and not both past and present, present and future, etc. than events are.

Nor does it help to distinguish the 'object language', in which events are said to be past, present, or future, from a 'meta-language' in which object-language sentences are said to be true or false. At least, it helps only if the meta-language sentences are the tenseless ones used to give tensed object-language sentences their tenseless token-reflexive truth-conditions, e.g. 'Tokens of "*t* is now" are true iff they are located at *t*'. When the meta-language sentences are themselves tensed, the problem simply reappears in a new guise. Truth and falsity are now the incompatible properties of the object-language sentence types (since tensed truth-conditions are supposed not to be token-reflexive; if they were, as we shall see later, they would not be tensed). But to say that a particular sentence type is true—i.e. that any token of it would be true—is to say that it is not false, and vice versa. Yet any true and non-trivially tensed sentence type will also be (sometime) false. This now is McTaggart's basic contradiction, and the riposte to it is the same: no tensed sentence type is both true and false at the same time. Meta-language sentences then say when these object-language sentences are true and when false. But if the meta-language sentences are themselves tensed, they too will be both true and false. The contradiction simply recurs in the meta-language. Removing it from there by using a tensed meta-meta-language to say when meta-language sentences are true and when false only leads to McTaggart's regress. Iterating tensed meta-languages no more refutes McTaggart than iterating tensed properties of events or facts does, or than iterating tensed operators on propositions or sentences within a single language.

The plain fact is that nothing can have mutually incompatible properties, whether they be tenses or truth-values: neither events, things, facts, propositions, sentences, nor anything else. I prefer therefore to stick to events and things, as being the natural inhabitants of *A* series locations. I will not translate the problem into other and more fashionable idioms, which only pander to the erroneous conviction that McTaggart can be thus easily answered.

Thirdly, however, I must deal with the complaint that in symbolizing McTaggart's argument I have myself begged the question against tenses. Specifically, in using 'P*e*' to say that *e* is past, I have left out the verb 'is'. By so doing I have tacitly treated the 'is' in '*e* is past' as a tenseless copula, which is why *e*'s being past, present, and future appears to be contradictory. For in fact the verb 'is' in '*e* is past' is tensed, i.e. it really means that *e* is *now* past. And, given that, the contradiction vanishes, since, if *e* is now past, it is not also now present or now future. Of course, it *was* future and it *was* present, but that is quite compatible with *e*'s being now past. In short, the supposed contradiction has been artificially generated by suppressing the essentially tensed verbs used in ascribing to *e* the properties P, N, and F.

This complaint misses the point of tense completely. The *A* series is supposed to be a feature of the world, not of verbs. We don't need tensed verbs to make tensed statements, i.e. statements ascribing *A* series locations to things and events. We can do that just as well by using adverbs and phrases

like 'yesterday', 'this week', and 'next year'. They make verbal tenses redundant, and so do the expressions 'in the past', 'now', and 'in the future',
i.e. 'P', 'N', and 'F'. That is their function: to take over the semantic roles
respectively of the past, present, and future verbal tenses to which by definition they are equivalent. Given these expressions, verbs might as well be
tenseless, i.e. take the same form regardless of the A series location of the
events they refer to. Suppose, for example, that 'happens' is made (by stipulation if necessary) into such a tenseless form. Then 'It happened' means 'It
happens in the past'. If the past-tense form of the verb in 'happened in the
past' were not redundant, that phrase would have to mean PP rather than P,
which it clearly doesn't. It simply means 'happened'. Just as 'in the past' is
superfluous given the past verbal tense, so the verbal tense is superfluous
given 'in the past'.

As for the past, so for the present. Adding 'now' to 'It happens' or 'It is
happening' makes the present-tense connotations of the verb superfluous.
Given the 'now', 'happens' is as tenseless in 'It happens now' as in 'It
happens in the past' and 'It happens in the future'. And as for 'happens', so
for 'is'. The temporal meanings of 'e is past', 'e is present', and 'e is future'
are supplied entirely by the words 'past', 'present', and 'future'. The 'is' *is*
a mere tenseless copula, present only because English grammar gives sentences verbs even when, as here, they contribute nothing to the content. Nothing,
therefore, is left out by abbreviating these sentences to 'Pe', 'Ne', and 'Fe'.
So the abbreviation can generate no contradiction that was not already there.

Anyone who still says the 'is' in 'e is past' is present tense, so that 'e is
past' means 'e is now past', will have to say what tense 'is' then has in 'e is
now past'. It is clearly either tenseless or present tense—and, if tenseless,
McTaggart's contradiction reappears at once, because 'e is now past' is not
always true. It cannot be, since it must be true when and only when 'e is past'
is true; otherwise those two sentences would not mean the same and the
one could not replace the other. So all we have to do to regenerate McTaggart's proof, as readers may again easily verify, is replace 'P', 'N', and 'F'
throughout by 'NP', 'NN', and 'NF'.

But if the 'is' in 'e is now past' is tensed, as in 'e is past', the same vicious
regress appears in the form of the verb itself. For 'e is past', meaning 'e is
now past', must now also mean 'e is now now past', in which again the 'is'
must either be tenseless or present tense. If tenseless, we again get McTaggart's argument, starting this time with 'NNP', 'NNN', and 'NNF'; and, if
present tense, the regress continues with 'e is now now now past', 'e is now
now now now past', and so on *ad infinitum*. And for no sentence type in this
endless sequence can we consistently give tensed truth-conditions. It is no
use saying, at any stage in the sequence, that the last sentence type in it is

true *now*, because whether that is so depends on when *now* is. Saying this merely generates the next type in the sequence, concerning which the same question arises. To stop and give a definite answer at any stage only produces a contradiction, because if the sentence is true (at some present time) it is also false (at some other). The only way to avoid contradiction is never to stop at all, which is tantamount to admitting that the original sentence type has no tensed truth-conditions, i.e. cannot be made true or false by any tensed fact such as that *e* is past, *e* is now past, *e* is now now past, etc. In short, supposing that there are such facts is either self-contradictory or useless for saying what makes tensed sentences true or false.

McTAGGART AND TOKEN-REFLEXIVES

So much by way of reinforcing McTaggart's own proof. But in case it still does not convince, I will now put it explicitly in terms of token-reflexive truth-conditions. First, we must note that tensed facts are supposed to provide *non*-token-reflexive truth-conditions for tensed thoughts and sentences. Just as all tokens of 'Snow is white', whenever and wherever they occur, are made true by the single fact that snow is white, so all true tokens of '*e* is past', whenever they occur, are supposed to be made true by the single fact that *e* is past—or, if for some reason that won't work, by the fact that *e* is now past, or that *e* is now now past, etc. All tokens of the same tensed type are supposed to have the same *tensed* truth-conditions, however much their tenseless truth-conditions may vary from token to token.

This looks possible because what prevents tensed thoughts and sentences from having non-token-reflexive tenseless truth-conditions is that tenseless facts, unlike tensed ones, are facts at all dates. That is why the tenseless truth-conditions of tokens of '*e* is past' must vary with their dates, to allow their truth-values to vary from 'false' before *e* to 'true' after it. But since the tensed facts that *e* is past, now past, now now past, etc. are *not* facts at all dates, but only at dates later than *e*, they should be able to give all tokens of '*e* is past' the same tensed truth-condition, namely, that *e* is past (or now past, or now now past . . .).

But they can't. None of these supposed tensed facts will give '*e* is past' correct non-token-reflexive truth-conditions, because even when they *are* facts, its tokens' truth-values still depend on their own dates. Thus at any date *t* later than *e*, the facts are that *e* is past, now past, now now past, etc. But none of these facts makes tokens of '*e* is past' which occurred before *e* true at *t*. Those tokens—as opposed to the *type* '*e* is past'—are false then, just as they always were and always will be. (A long-lasting token can of course vary in truth-value during its lifetime, e.g. a token of '*e* is past' printed before *e* will start off true and end up false. But that does not change the truth-value it had earlier—any more than a death posthumously verifies premature announcements of it.)

But perhaps these tensed facts will give '*e* is past' correct non-token-reflexive truth-conditions when its tokens' temporal locations are specified in *A* series terms? Unfortunately for them, they won't. For even when it is *now* a fact that *e* is past, *e* is now past, etc., tokens of '*e* is past' are still not all true now regardless of their own *A* series locations. Tokens that are now more past than *e* itself will be false now, just as they always were and always will be. So even when stated entirely in *A* series terms the tensed truth-conditions of '*e* is past' remain token-reflexive, i.e. they vary with its tokens' *A* series locations.

And as for '*e* is past', so for all seriously tensed thoughts and sentences, i.e. those whose tokens would be true at some dates and false at others. (For present purposes we may ignore tensed thoughts and sentences made contingently false at all dates by non-temporal facts, e.g. 'Napoleon won at Waterloo last year'. The fact that Napoleon lost is irrelevant to this sentence's temporal meaning: what matters here is how its tokens' truth-values would have varied with time if he'd won.) Some pairs of tokens of any such tensed type will therefore differ in truth-value just because their *B* series locations differ. But whatever has a *B* series location also has, at any *B* series moment, a corresponding *A* series location. So, in particular, these pairs of tensed tokens will also always differ in their *A* series locations, with which their truth-values will therefore also always vary. In short, if the tenseless truth-conditions of tensed thoughts and sentences are token-reflexive, so must their tensed truth-conditions be.

It follows that trying to give tensed thoughts or sentences non-token-reflexive truth-conditions, tensed or tenseless, always leads to contradiction. That, for tenseless truth-conditions, is what shows why tensed sentences cannot be translated by tenseless ones: since all tokens of any such translation would have to have the same truth-value regardless of their date, they will contradict some possible tokens of the original sentence, whose truth-values do vary from date to date. And the same goes for tensed truth-conditions. Since the truth-value of tokens of any such tensed sentence will also vary with their *A* series locations, our giving them all the same truth-value because some tensed truth-condition does or doesn't obtain *now* will inevitably conflict with the truth-value of some of the sentence's past or future tokens.

This, in token-reflexive terms, is McTaggart's contradiction. That it is so is most easily seen in the meta-language version of his argument given above. Because the tensed truth-conditions of '*e* is past' are token-reflexive, any attempt to state in a tensed meta-language the one tensed fact that makes all its true tokens true is bound to fail. The alleged fact would by definition have to make all tokens of the type true, regardless of their A series location, whereas in fact some are always true and others always false. Hence the contradiction. And it is, I hope, easier to see in this version of the argument that complex tenses are no better off, i.e. that the regress of meta-languages McTaggart's critics invoke is indeed vicious. For the above argument applies to tenses of any complexity. Provided only that, as everyone now agrees, all thoughts and sentences of all tensed types have tenseless token-reflexive truth-conditions, their tensed truth-conditions will also be token-reflexive. And whatever doubts there may be about the token-reflexivity of some complex tenses, there can be none about those McTaggart's critics resort to. For, as I have already remarked, unless 'e is now past', 'e is now now past', etc. had the same token-reflexive truth-conditions as 'e is past', they could not be substituted for it. And if they do have those truth-conditions, then McTaggart's argument disposes of the tensed facts they allegedly state, just as it disposes of the alleged fact that e is past.

Finally, I suppose defenders of tense might ask why tensed truth-conditions cannot be token-reflexive, if tenseless ones are. The answer is that they then cease to be tensed. Suppose, for example, it is now n years after e: e is now past, but that fact alone does not suffice to make all tokens of 'e is past' now true. However, consider a token that is only m years past, where m is less than n. The token is true, because it is less past than e itself. Those are the ostensibly tensed truth-conditions of any token of the type: it is true if and only if when it is present e is past, i.e. if and only if $n - m$ is positive. But if $n - m$ is ever positive, it is always positive, because the temporal distance between A series locations never changes. The values of n and m continually increase, but always at the same rate, so the value of their difference stays the same. The fact is simply that the token is—tenselessly— $n - m$ years later than e. The variably tensed elements n and m in the supposedly tensed token-reflexive truth-conditions cancel out, leaving the already familiar tenseless truth-conditions: true if later than e, false otherwise.

Similarly for all other tensed sentence types. Their tensed truth-conditions are either self-contradictory or token-reflexive; and, if token-reflexive, they reduce to tenseless truth-conditions. As McTaggart saw, the truth-conditions of tensed sentences are either tenseless or self-contradictory.

My version of McTaggart's proof started from the fact that all tensed sentences and judgements have tenseless token-reflexive truth-conditions. To start with I left open the possibility that tensed sentences also state tensed facts;

but we can see now that this is not a real option after all. And while those who have immunized themselves against McTaggart's proof of the unreality of tense may need something like the above argument to convince them, here finally is a much quicker argument which should serve to sway more open minds.

The sole function of tensed facts is to make tensed sentences and judgements true or false. But that job is already done by the tenseless facts that fix the truth-values of all tokens of tensed thoughts and sentences. Provided a token of 'e is past' is later than e, it is true. Nothing else about e and it matters a jot; in particular no tensed fact about them matters. It is immaterial, for a start, where e and the token are in the A series; and if that is not material, no more recherché tensed fact can be. Similarly for tokens of all other tensed types. Their tenseless truth-conditions leave tensed facts no scope for determining their truth-values. But these facts by definition determine their truth-values. So in reality there are no such facts.

2

RELATIONISM ABOUT TIME

IV

TIME WITHOUT CHANGE

SYDNEY SHOEMAKER

It is a widely held view that the passage of time necessarily involves change in such a way that there cannot be an interval of time in which no changes whatever occur. Aristotle spoke of time as 'a kind of affection of motion', and said that, although time cannot be simply equated with motion or with change, 'neither does time exist without change'.[1] Hume claimed that ' 'tis impossible to conceive . . . a time when there was no succession or change in any real existence'.[2] And McTaggart presented as something 'universally admitted' the contention that 'there could be no time if nothing changed' (from which, he claimed, it follows that everything is always changing, at least in its relational qualities).[3] Similar claims can be found in the works of contemporary writers.[4]

The claim that time involves change must of course be distinguished from the truism that change involves time. And, as it will be understood in this paper, it must also be distinguished from a truism that Aristotle expressed by saying 'if the "now" were not different but one and the same, there would not have been time', i.e. the truism that if at time t' some time has elapsed since time t, then t' is a different time than t.[5] I do not think that this truism is what Aristotle had in mind in asserting that time involves change, but it, and certain related truisms, have seemed to some philosophers, e.g. to McTaggart, to imply that there are changes that occur with a logically

Sydney Shoemaker, 'Time without Change', *Journal of Philosophy*, 66 (1969): 363–81. Reprinted by permission of the author and the *Journal of Philosophy*. The text as reprinted here includes some small changes made by the author when this essay was reprinted in his *Identity, Cause, and Mind* (Cambridge: Cambridge University Press, 1984).

[1] *Physics*, IV, ch 11, 218[b].

[2] *A Treatise of Human Nature*, ed. L. A. Selby-Bigge (Oxford: Oxford University Press, 1888), 40.

[3] J. M. E. McTaggart, *The Nature of Existence* (Cambridge: Cambridge University Press, 1927), ii. 11.

[4] See e.g. Bruce Aune's 'Fatalism and Professor Taylor', *Philosophical Review*, 71(4) (Oct. 1962): 512–19: 518, and, for a somewhat more qualified statement of the view, Jonathan Bennett's *Kant's Analytic* (New York: Cambridge University Press, 1966), 175. Bennett makes the acute point that, because of the multi-dimensionality of space and the unidimensionality of time, empty space is measurable in ways in which empty time necessarily is not.

[5] *Physics*, IV, ch. 11, 218[b]

necessary inevitability and relentlessness. Thus the date and time of day is constantly changing, it is constantly becoming later and later, whatever exists is constantly becoming older and older (whether or not it 'shows its age'), and not a moment goes by without something that had been future becoming present and something that had been present becoming past. Such changes, if indeed they are changes, are bound to occur no matter how much things remain the same; whatever else happens or fails to happen in the next twenty-four hours, the death of Queen Anne (to use Broad's example) is bound to recede another day into the past.

I do not wish to become embroiled, in this paper, in the controversy as to whether these 'McTaggartian' changes deserve to be regarded as genuine changes. My own view is that they do not. But my concern in this paper is with ordinary becoming, not 'pure becoming'; my concern is with changes with respect to such properties as colour, size, shape, weight, etc., i.e. properties with respect to which something *can* remain *un*changed for any length of time. And, though McTaggart may be an exception, I think that philosophers who have claimed that time involves change have generally meant, not of course that everything must always be changing with respect to every such non-McTaggartian property, but that during every interval of time, no matter how short, something or other must change with respect to some such property or other.

This view, unlike the truism that time involves McTaggartian change, has important cosmological consequences. It implies, for example, that the universe cannot have had a temporal beginning unless time itself had a beginning and that the universe cannot come to an end unless time itself can come to an end. The claim that time involves McTaggartian change is compatible with the universe having had a beginning preceded by an infinite span of empty time, for throughout such a span the beginning of the universe, and the various events in its history, would have been 'moving' from the remote future toward the present, and this itself would be McTaggartian change. But the kinds of change I am here concerned with are changes of things or substances, not of events, and such a change can occur only while the subject of change exists; the occurrence of such changes involves the existence of a universe of things, and if time involves change then there can be no time during which the universe does not exist.

There is another sort of change, or ostensible change, which must be ruled out of consideration if the claim that time involves change is to assert more than a triviality. Consider Nelson Goodman's term 'grue', and suppose that this is given the following definition (which, though not Goodman's definition, is common in the literature): 'x is grue at t if and only if t is earlier than AD 2000 and x is green at t or t is AD 2000 or later and x is blue at t'. Anything that is green up to AD 2000 and remains green for some time after AD 2000 necessarily changes at AD 2000 from being grue to being non-grue. Clearly, for any interval during which something remains unchanged with respect to any property whatever, we can invent a 'grue'-like predicate which that thing either comes to exemplify or ceases to exemplify during that interval. And if we take there to be a genuine property corresponding to every grue-like predicate and count the acquisition or loss of such properties as genuine change, it follows that whenever anything remains unchanged in any respect it changes in some other respect. Now it is notoriously difficult to justify or explicate the intuition that there is a distinction between greenness and grueness which justifies regarding the former but not the latter as a genuine property, and it is correspondingly difficult to justify or explicate the intuition that something does not undergo a genuine change when at the advent of the year AD 2000 it ceases to be grue by continuing to be green. But I shall assume in this paper that these intuitions are well founded and shall exclude from consideration 'changes' that, intuitively, consist in the acquisition or loss of 'positional', i.e. grue-like, qualities. If we do not do this, the view that time involves change becomes trivially and uninterestingly true, and the considerations usually advanced in its favour become irrelevant to it.

Aristotle's statement of his grounds for thinking that time involves change is unclear but suggestive. He says that

> when the state of our own minds does not change at all, or we have not noticed its changing, we do not realize that time has elapsed, any more than those who are fabled to sleep among the heroes in Sardinia do when they are awakened; for they connect the earlier 'now' with the later and make them one, cutting out the interval because of their failure to notice it. (Ibid.)

It is not clear to me why Aristotle focuses here on change of 'the state of our own minds', although later on I shall venture a suggestion about this. But if we leave this aside, the argument seems to be that time involves change because the awareness, or realization, that an interval of time has elapsed necessarily involves the awareness of changes occurring during the interval. It is not a serious objection to this that sometimes, e.g. when we have been asleep, we are prepared to allow that a good deal of time has elapsed since a given event occurred even though we were not ourselves aware of any changes during the interval, for in such cases it is plausible to hold that our belief that an interval of a certain duration has elapsed is founded on the inductively grounded belief that changes did occur that we could have been aware of had we been awake and suitably situated.

What Aristotle says here seems to be supported by the obvious and often mentioned fact that it is by observing certain sorts of changes, e.g. the movements of clock hands, pendulums, and the sun and stars, that we measure time. Even if what we are measuring is the length of time during which a given object remained *un*changed, it seems necessary that something, namely whatever we are using as our clock, should have changed during that interval. This is perhaps what Aristotle meant when he said that time is directly the measure of motion and only indirectly the measure of rest. At any rate, the fact that we measure time by observing changes lends plausibility to the view that there cannot be an interval of time in which no changes occur. The contrary view can seem to lead to total scepticism about the possibility of measuring time. If it is possible for there to be changeless intervals, then it may seem compatible with my total experience that any number of such intervals, each of them lasting billions of years, should have elapsed since I ate my last meal, despite the fact that the hour hand of my watch has made only one revolution and the fact that my lunch is still being digested. For if such intervals can occur there is apparently no way in which we can be assured of their non-occurrence; as Aristotle put it, 'the non-realization of the existence of time happens to us when we do not distinguish any change' (ibid.). And if this is so, we can never know how much time has elapsed since the occurrence of any given past event. But, it may be held, if the supposition that changeless intervals are possible leads to this sort of scepticism, this itself is proof that the supposition is false.

Of course, it is not only by measurement, i.e. by the use of clocks and the like, that we are aware of the existence and extent of intervals of time. We are all possessed of a 'sense of time', an ability to judge fairly accurately the length of intervals of time, at least of short intervals, without using any observed change as a standard; one can tell whether the second hand of a clock is slowing down without comparing its movements with those of another clock, and if one hears three sounds in succession one can often tell without the aid of a clock or metronome how the length of the interval between the first and second compares with that of the interval between the second and third. But, although the exercise of this ability to judge the length of temporal intervals need not involve *observing* any change, it is plausible to suppose that as long as one is aware of the passage of time some change must be occurring, namely, at a minimum, a change in one's own cognitive state. Suppose that throughout an interval of five minutes I observe just one object, call it O, which remains completely unchanged throughout the interval, and that at each point during the interval I know how long I have observed O to remain unchanged. Then the content of my knowledge will be different at different moments during the interval. For example, at one time I will know that I have been observing O for two minutes, and a minute later I will know that I have been observing O for three minutes. And this means that there will be a constant change in my cognitive state as the interval progresses.[6] Possibly it was considerations of this sort that led Aristotle to stress change 'in the state of our own minds' in his discussion of the relationship between time and change—although it does not seem to be true to say, as he does, that one must *notice* a change in the state of one's own mind in order to be aware of the passage of time.

These considerations suggest that it is logically impossible for someone to know that nothing, including the state of his own mind, is changing, i.e. for someone to be aware of the existence of a changeless interval during that interval itself. But it does not of course follow from this that it is impossible for someone to be aware of the existence of such an interval before or after its occurrence. To take an analogous case, it is logically impossible that anyone should know, at any given time, that the then current state of the universe is such as to make impossible the existence in it of life and consciousness, yet most of us believe that we have very good reasons for thinking that the universe has been in the very remote past, and will again be in the very remote future, in just such a state. In what follows I shall try to show that it is conceivable that people should have very good reasons for thinking that there are changeless intervals, that they should have well grounded beliefs about when in the past such intervals have occurred and when in the future they will occur again, and that they should be able to say how long such intervals have lasted or will last. Of course, the fact that people might have good reasons for thinking that something happens does not prove that it is logically possible for that thing to happen; people have had good reasons for thinking that the circle has been squared. But I think that the sorts of

---

[6] Suppose, however, that what I am aware of, at each moment during the interval (after the first minute of it), is only that O has remained unchanged—has remained in a certain state which I will call S—during the immediately preceding minute. (I have, let us suppose, an incredibly short memory span, and after the first minute of the interval my memory does not extend back to the beginning of it.) Would this continuous awareness of lack of change in O involve a continuous change in my own state of mind? One might argue that it does, on the grounds that at each instant I know something I did not previously know, namely that at *that* instant O is, and has been continuously for one minute, in state S. On the other hand, one could argue that my cognitive state at any instant during the interval (after the first minute) consists in a certain predicate's being true of me, namely the predicate 'knows that O has remained in state S for the last minute', and that since the very *same* predicate is true of me throughout there is no change. I shall not here try to resolve the tricky issue of which of these ways of viewing the matter is correct. I shall only remark that the former, according to which awareness, even of changelessness, involves change on the part of the subject of awareness, seems to me essentially the same as C. D. Broad's view that as long as one is conscious there is a 'steady movement of the quality of presentedness' along the series of one's experiences; see his *An Examination of McTaggart's Philosophy*, vol. ii, pt. 1 (Cambridge: Cambridge University Press, 1938), 308. My present inclination is to regard this kind of 'change' as a species of McTaggartian pseudo-change. The issues raised by this example are similar to those raised by a very interesting argument of Norman Kretzmann's, to the effect that God must always be changing if he always knows what time it is, and that there is therefore an incompatibility between the claim that God is omniscient and the claim that he is immutable. See Kretzmann's 'Omniscience and Immutability', *Journal of Philosophy*, 63(14) (14 July 1966): 409–21

grounds there could conceivably be for believing in the existence of change-less intervals are such that no sound argument against the possibility of such intervals can be built on a consideration of how time is measured and of how we are aware of the passage of time.

To the best of my knowledge, it follows from well-established principles of physics that our universe is a perpetually changing one. But what is in question here is not whether it is physically possible for there to be time without change but whether this is logically or conceptually possible. Accordingly, I shall allow myself in what follows to consider 'possible worlds' in which the physical laws differ drastically from those which obtain in the actual world. It may be objected that scientific progress brings con-ceptual change and that within modern physical theory it is not possible to make any sharp distinction between those propositions about time which express logical, or conceptual, claims and those which purport to express synthetic truths of physics. But I think that it is fair to say that those philo-sophers who have claimed that time involves change have not generally rested their case on recent developments in physics, e.g. relativity theory, and have thought that this claim holds for our ordinary, pre-scientific, concept of time as well as for the more sophisticated conceptions provided by the physicists. And in dealing with such a view it seems to me legitimate to consider possible worlds in which quite different physical theories would be called for. If someone wishes to maintain that the occupants of such a world would necessarily have a different concept of time than that which the physicists tell us is applicable to our world, I have no objection to make—as long as it is granted that their concept would have enough in common with our notion of time to make it legitimate to regard it as a concept of *time*. I should concede that in allowing myself to speak of worlds that are logically but not physically possible I am making the somewhat controversial assumption that there is a tenable distinction between logically contingent and logically necessary truths. But this assumption is one that I share with the philosophers against whom I am arguing, those who say that time involves change—for I think that this claim is philosophically interesting only if we understand the 'involves' in it as meaning 'necessarily involves'.

Consider, then, the following world. To the best of the knowledge of the inhabitants of this world all of its matter is contained in three relatively small regions, which I shall call A, B, and C. These regions are separated by natural boundaries, but it is possible, usually, for the inhabitants of this world to pass back and forth from one region to another, and it is possible for much of what occurs in any of the regions to be seen by observers situated in the other regions. Periodically there is observed to occur in this world a phenomenon which I shall call a 'local freeze'. During a local freeze all processes occurring in one of the three regions come to a complete halt; there is no motion, no growth, no decay, and so on. At least this is how it appears to observers in the other regions. During a local freeze it is impossible for people from other regions to pass into the region where the freeze exists, but when inhabitants of other regions enter it immediately following the end of a freeze they find that everything is as it would have been if the period of the freeze had not occurred. Eggs laid just prior to the beginning of a freeze lasting a year are found to be perfectly fresh; a glass of beer drawn just prior to the beginning of the freeze still has its head of foam, and so forth. And this remains so even when they make the finest measurements, and the most sophisticated tests, available to them; even radioactive decay, if such exists in this world, is found to be completely arrested during the period of a local freeze. Those people who were in the region during the freeze will initially be completely unaware that the period of the freeze has elapsed, unless at the beginning of the freeze they happened to be observing one of the other regions. A man who was stopped in the middle of a sentence by the onset of the freeze will resume the sentence at the end of it, and neither he nor his hearers will be aware that there has been any interruption. However, things will seem out of the ordinary to any inhabitant of a frozen region who at the beginning of the freeze was looking into one of the other regions. To such a person it will appear as if all sorts of major changes have occurred instantaneously in the other region: people and objects will appear to have moved in a discontinuous manner or to have vanished into thin air or to have materialized out of thin air; saplings will appear to have grown instantaneously into mature trees; and so on. Although people might initially refuse to believe that events that seem to them to have only just occurred in fact occurred a year before and that they have been unconscious for a full year, it would seem that they would eventually come to believe this after hearing the reports of observers from other regions and, more important, after they themselves have observed local freezes in other regions.

The possibility of what I have described so far is compatible with the claim that there can be no time without change. That claim is that *something or other* must change during any interval of time and not that everything must change during every interval, and all that I have so far described is a case in which a fairly large percentage of the things in my imaginary world remain unchanged (or apparently unchanged) throughout an interval of time. But now the following seems possible. We can imagine, first, that the inhabitants of this world discover, by the use of clocks located in unfrozen regions, that local freezes always last the same amount of time—let us suppose that the length of freezes is always exactly one year. We can also imagine that they keep records of local freezes and find that they occur at regular intervals—let us suppose that it is found that in region A local freezes have occurred every third year, that in region B local freezes have occurred every fourth year, and that in region C local freezes have occurred every fifth year. Having noticed this they could easily calculate that, given these frequencies, there should be simultaneous local freezes in regions A and B every twelfth year, in regions A and C every fifteenth year, in regions B and C every twentieth year, and in all three regions every sixtieth year. Since these three regions exhaust their universe, to say that there will be simultaneous local freezes in all three regions every sixtieth year is to say that every sixtieth year there will be a *total* freeze lasting one year. Let us suppose that the predicted simultaneous two-region freezes are observed to occur as scheduled (the observers being, in each case, the inhabitants of whichever region remains unfrozen), that no freeze is observed to begin by anyone at the time at which local freezes are scheduled to begin simultaneously in all three regions, and that the subsequent pattern of freezes is found to be in accord with the original generalization about the frequency of freezes. If all of this happened, I submit, the inhabitants of this world would have grounds for believing that there are intervals during which no changes occur anywhere.[7]

The objections that might be made to this (and they are many) can be divided into two sorts. Objections of the first sort maintain, on various grounds, that the inhabitants of my imaginary world could not really have good reasons for believing that no changes whatever occur in a region during an ostensible local freeze in that region. For example, it might be held that, even if the hypothesis that no changes occur in such regions has survived a large number of refinements of their instruments and techniques of measurement, they could never be entitled to believe that further refinements of their instruments and techniques would not show that very slight changes occur during such intervals. Or it might be held that visual observation of an ostensibly frozen region would itself involve the occurrence of changes in that region, namely the transmission of light rays or photons. Objections of the second sort do not question the possibility of there being good reasons for believing in the occurrence of local freezes, but do question the legitimacy of extrapolating from these to the periodic occurrence of total freezes. Later on two objections of this sort will be considered in detail.

I shall not in this paper consider, except in a very general way, objections of the first sort. For though I am inclined to think that all such objections can

---

[7] It is obvious that during a local freeze objects in the frozen region will undergo changes of a kind; they will undergo changes in their relational properties in virtue of the changes that are still going on in the unfrozen regions. But during a total freeze there are no unfrozen regions, and so no changes occur even with respect to relational properties.

be met, I think that such objections have limited force even if correct.[8] Even if the inhabitants of this world could not have good grounds for thinking there are intervals in which no changes at all occur, it seems clear that they could have good grounds for thinking there are intervals in which no changes occur that are detectable by available techniques and instruments. And this goes against the view, suggested by Aristotle's remarks, that when we have the well-grounded belief that two events are separated by an interval of time this belief is always grounded, ultimately, on evidence that changes occurred between these events, i.e. is grounded either on observations of such changes or on inductive evidence that such changes occurred. Moreover, if one thinks that the possibility of time without change can be ruled out on verificationist grounds and if it is only objections of the first sort that enable one to maintain that it is impossible to verify the existence of changeless intervals, then one seems to be committed to a view which is much stronger, and intuitively less plausible, than the view that *something or other* must change during every interval of time; one seems committed to the view that *everything* must change during every interval of time. Now there is of course a sense in which a change in any given thing involves a change in the relational properties of everything else. But it now appears that the verificationist must rest his case on the (alleged) impossibility of verifying that anything has remained wholly unchanged even with respect to its *non*-relational properties and that he ought to conclude that it is logically impossible for anything to remain unchanged with respect to its non-relational properties. But this seems no more plausible than the argument from the fact (if it is one) that it is impossible to verify that two things are exactly equal in length to the conclusion that any two things necessarily differ in length.

I turn now to objections of the second sort, and to the first objection that I shall consider in any detail. I have imagined that the inhabitants of my imaginary world come to accept the generalization that local freezes occur in region A every three years, in region B every four years, and in region C every five years, from which it follows that there is a total freeze every sixty years. But why should they accept this generalization? What they observe is equally compatible with the generalization that freezes occur with these

---

[8] Of the two objections of this sort I have mentioned, I think that the first can be met by supposing that the scientific investigations of these people support a 'quantum' theory of change which rules out the possibility of changes so slight that they are undetectable by certain instruments. The second could be met by supposing that visual observation in this world does not involve the occurrence of processes in the vicinity of the thing perceived, does not involve the transmission at finite velocities of waves or particles. Alternatively, we can avoid the objection by supposing that while a local freeze exists in a region it is as if the region were divided from the rest of the world by an opaque (and impenetrable) curtain, and that what serves as evidence that no change occurs in regions thus insulated is the fact that when such a region again becomes observable everything appears to be just as it was immediately before the region became insulated

frequencies *with the exception* that all three regions skip a freeze every fifty-nine years; or in other words (to put this in a way that makes it sound less *ad hoc*): one-year local freezes occur in A in cycles in which nineteen freezes occur at the rate of one every third year, with four 'freeze-less' years between the last freeze of one cycle and the first freeze of the next; they occur in B in cycles in which fourteen freezes occur at the rate of one every fourth year, with six years between cycles; and they occur in C in cycles in which eleven freezes occur at the rate of one every fifth year, with eight years between cycles. This generalization does not imply that there are ever freezes in all three regions at the same time, and it may be held that for just this reason it should be preferred to the generalization that does imply this.

Now it seems to be generally agreed that if two hypotheses are compatible with the same observed data, we should prefer the simpler of the hypotheses in the absence of a good reason for preferring the other. And the first generalization stated above seems clearly simpler than the second. One reason for preferring the second is the belief that total freezes, i.e. changeless intervals, are impossible. And the most common basis for this belief is the conviction that the existence of changeless intervals is unverifiable. But, on the assumption that the simpler hypothesis is a possible one, the existence of total freezes is verifiable by standard inductive procedures; so one cannot claim that the existence of changeless intervals is unverifiable without begging the question against the possibility of the simpler hypothesis. Of course, the existence of total freezes is not 'directly' verifiable, if direct verification of the occurrence of something involves knowing of its occurrence while it is actually occurring. But there are all sorts of things whose occurrence is not directly verifiable in this sense and yet is perfectly possible and knowable; it would be impossible to verify directly, in this sense, that the rotation of the earth would continue if everyone in the universe were sound asleep, yet it is clearly possible that everyone in the universe should at some time be sound asleep, and we all have excellent reasons for believing that if this ever happens the rotation of the earth will continue. I conclude that considerations of verification give no reason for preferring the second hypothesis to the first, and that the first, being simpler, should be preferred unless some other reason for preferring the second can be found.

If one does not find this wholly convincing, this is probably because the generalization that implies the existence of total freezes does not strike one as significantly simpler than its competitor, and because one views the latter not really as a 'hypothesis' at all, but rather as a straightforward description of what would actually be observed over a long period of time by the inhabitants of my imaginary world. But I think that this way of viewing the matter becomes less plausible if we introduce some modifications into the example.

So far I have supposed that local freezes are always of the same length, and that whenever local freezes in different regions coincide they do so completely, i.e. begin and end at the same times. Let us now suppose instead that freezes vary in length and that sometimes freezes in two different regions overlap, so that the inhabitants of each region can observe part of the freeze in the other region, namely the part that does not coincide with a freeze in their own region. Let us further suppose that the length of local freezes is found to be correlated with other features of the world. For example, we can suppose that immediately prior to the beginning of a local freeze there is a period of 'sluggishness' during which the inhabitants of the region find that it takes more than the usual amount of effort for them to move the limbs of their bodies, and we can suppose that the length of this period of sluggishness is found to be correlated with the length of the freeze. Finally, let us replace the supposition that observed freezes always last one year with the supposition that they always last longer than six months.

It now becomes possible to decide empirically between the two hypotheses stated earlier. First, it is compatible with the first and simpler hypothesis, but not with the second, that during the sixtieth year after the beginning of a cycle some periods of freeze should be observed. For now we are allowing local freezes to overlap and to last for less than a full year, and this allows freezes to be observed even in a year in which there are freezes in all three regions. Perhaps the second hypothesis could be modified in such a way as to allow there to be local freezes during the sixtieth year, as long as there is no interval during which all three regions are simultaneously frozen. This would of course involve asserting that there are exceptions to the rule that freezes always last longer than six months. Moreover, it obviously could turn out that on occasions on which the local freezes in the sixtieth year could not have lasted longer than, say, four months without there having been a period of total freeze, the periods of sluggishness preceding them were observed to be of a length that had been found in other cases to be correlated with freezes lasting, say, seven months. We can of course modify the second hypothesis still further, so that it will assert that there are exceptions to the rule that the length of the freeze is always proportional to the length of the adjacent period of sluggishness, and that these exceptions occur every fifty-nine or sixty years. But this does seem to me to make the hypothesis patently *ad hoc*. By positing these sorts of exceptions to observed regularities one can of course make the second hypothesis compatible with the observed facts, but it seems to me that this is no more intellectually respectable than the use of the same procedure to protect from empirical falsification the quasi-Berkeleian hypothesis that objects disappear when no one is looking at them, or, to take a case closer to home, the hypothesis that it is impossible for there to be an interval of time during which everyone in the world is sound asleep.

This brings me to the last objection that I shall consider. Suppose for the moment that it is correct to describe my imaginary world as one in which there are intervals during which no changes, and hence no events or processes, occur. A question arises as to how, in such a world, processes could get started again after the end of such an interval, i.e. how a total freeze could come to an end. What could *cause* the first changes that occur after there has been a total freeze? In the case of local freezes we might initially suppose that the end of a freeze, i.e. the changes that mark its termination, are caused by immediately preceding events (changes) in regions adjoining the region in which the freeze existed. But we cannot suppose that local freezes are terminated in this way if we want to defend the legitimacy of extrapolating from the frequency of their occurrence to the periodic occurrence of total freezes. For such an extrapolation to be legitimate, we must think of a total freeze as consisting in the simultaneous occurrence of a number of local freezes, the beginnings and endings of which are caused in the same way as are those of the local freezes from which the extrapolation is made. And if a freeze is total, there is no 'unfrozen' region adjoining any frozen region, and hence there is no possibility that the end of the freeze in any such region is caused by an immediately preceding event in an adjoining region. If there were evidence that the changes that terminate local freezes are always caused by immediately preceding events in adjoining regions, this would be a reason for rejecting the extrapolation to the existence of total freezes of fixed and finite durations. Nor does it seem open to a defender of the possibility of total freezes to hold that the changes that terminate freezes are uncaused events. For if that were so, it would apparently have to be sheer coincidence that observed freezes always last exactly one year (or, in the modified version of the example, that their length is proportional to that of the temporally adjoining intervals of sluggishness)—and it is illegitimate to extrapolate from an observed uniformity that one admits to be coincidental. So we are faced with the question: by what, if not by an immediately preceding event in an adjoining unfrozen region, could the end of any freeze be caused? And a special case of this is the question of how the end of a total freeze could be caused.

If we make the simplifying assumption that time is discrete, i.e. that for any instant there is a next instant and an immediately preceding instant, it is clear that the cause of the change that ends a total freeze cannot be, and cannot be part of, the state of the world in the immediately preceding instant. For the immediately preceding instant will have occurred during the freeze (will have been the last instant of the freeze), and since no change occurs during a total freeze the state of the world at that instant will be the same as its state at any other instant during the freeze, including the first one. If the state of the world at that instant were causally sufficient to produce a generically different world state in the immediately following instant, then the freeze would not have occurred at all, for then the change that ends the freeze would have begun immediately after the first instant of the freeze—and a freeze 'lasting' only an instant would be no freeze at all.

If time is dense or continuous, of course, we cannot in any case speak of a change as being caused by the state of the world at the immediately preceding instant, for in that case there is no immediately preceding instant. But I think that it is rather commonly supposed that if an event E occurs at time t and is caused, then, for any interval i, no matter how short, that begins at some time prior to t and includes all the instants between that time and t, the sequence of world states that exist during i contains a sufficient cause of E. If this is so, however, the first change that occurs after a total freeze could not have a cause. For let i be an interval with a duration of one second. If the freeze lasted more than one second, then the sequence of states that occurred during i was part of the freeze, and consequently the very same sequence of states occurred during the first second of the freeze. If the occurrence of that sequence of states had been sufficient to initiate the change that ended the freeze, the freeze could not have lasted more than one second. But since we can let i be as small an interval as we like, we can show that if the change that ends the freeze was caused, then the duration of the freeze was shorter than any assignable length, and this is to say that no freeze occurred at all.

It would seem that the only alternative to the view that the termination of a total freeze cannot be caused is the view that there can be a kind of causality that might be called 'action at a temporal distance' and that the mere passage of time itself can have causal efficacy. To hold this is to deny the principle, stated above, that if an event is caused then any temporal interval immediately preceding it, no matter how short, contains a sufficient cause of its occurrence. I shall refer to this principle as P. To suppose P false is to suppose that an event might be caused directly, and not via a mediating causal chain, by an event that occurred a year earlier, or that an event might be caused by such and such's having been the case for a period of one year, where this does not mean that it was caused by the final stage of a process lasting one year. Now I think that we are in fact unwilling to accept the existence of this sort of causality in our dealings with the actual world. If we found that a flash is always followed, after an interval of ten minutes, by a bang, we would never be willing to say that the flashes were the immediate causes of the bangs; we would look for some kind of spatio-temporally continuous causal chain connecting flashes and bangs, and would not be content until we had found one. And if we found that things always explode after having been red for an hour, we would never suppose that what causes the explosion is simply a thing's having been red for an hour; we would assume that there must be some process occurring in something that is red, e.g. the burning of a fuse or the uncoiling of a spring or the building up of an electric charge, and that the explosion occurs as the culmination of this process.

In the *Treatise* (though not in the *Inquiry*) Hume made it part of his definition of 'cause' that causes are 'contiguous' with their effects. And I think that there is some temptation to think that principle P, which could be thought of as expressing (among other things) the requirement that causes and their effects be temporally contiguous, is an analytic or conceptual truth. Establishing that this is so would not show directly that it is not logically possible for there to be changeless intervals, but it would undermine my strategy for arguing that this is logically possible. For, as we have seen, this would make it illegitimate for the inhabitants of my imaginary world to argue for the existence of total freezes on the basis of the observed frequency of local freezes.

But is P analytically or conceptually true? Here it is useful to distinguish two ostensible sorts of 'action at a temporal distance', both of which are ruled out by P. The first might be called 'delayed-action causality', and would be possible if the following were possible: X's happening at t is causally sufficient for Y's happening at a subsequent time t', and is compatible with t and t' being separated by an interval during which nothing happens that is sufficient for the occurrence of Y at t'. If in my earlier example we deny that the flash can be the 'direct' cause of the bang, we are denying that this sort of causality is operating. I think that it is commonly believed that this sort of causality is logically impossible, and I am inclined to believe this myself. But in order to save the intelligibility of my freeze example we do not need to assume the possibility of this extreme sort of causality at a temporal distance. All that we need to assume is the possibility of the following: X's happening at t is a necessary but not sufficient part of an actually obtaining sufficient condition for Y's happening at t', and t and t' are separated by an interval during which nothing happens that is sufficient for Y's happening at t'. To posit this sort of causality is not necessarily to deny the principle that causes must be temporally contiguous with their effects. If we take something's exploding at t to be the result of its having been red for the preceding hour, there is a sense in which the cause (the thing's having been red for an hour) is temporally contiguous with the effect (the explosion); yet here the thing's having been red at t-minus-one-half-hour is taken to be a necessary though insufficient part of a sufficient condition of its exploding at t, and it is assumed that nothing that happens during the intervening half-hour is sufficient to bring about the explosion. Likewise, if S is the state the world is in at every instant during a given total freeze and if E is the event (the change) that terminates the freeze, we can suppose that E is caused by the world's having been in state S for one year without violating the principle that causes are temporally contiguous with their effects, although not without violating principle P. Now we are, as I have already said, quite unwilling to believe that this sort of causality ever occurs in our world. But I am unable to see any conceptual reason why it could not be reasonable for the inhabitants of a world very different from ours to believe that such causality does occur in their world, and so to reject any principle, such as P, which excludes the possibility of such causality. And if this is possible, then in such a world there could, I think, be strong reasons for believing in the existence of changeless intervals.[9]

But here an important reservation must be made. Early in this paper I ruled out of consideration what I called 'McTaggartian changes', and in doing so I was implicitly refusing to count certain predicates, e.g. 'present' and '10 years old', as designating genuine properties—these (which I will call 'McTaggartian predicates') are predicates something comes to exemplify or ceases to exemplify simply in virtue of the passage of time. In ruling such predicates, and also grue-like predicates, out of consideration, I relied on what seem to be widely shared intuitions as to what are and what are not 'genuine' changes and properties. But these intuitions become somewhat cloudy if we try to apply them to a world in which there is action (or causal efficacy) at a temporal distance. Supposing 'F' to be a non-McTaggartian predicate, let us define the predicate 'F'' as follows: 'x is F' at t' $=_{df}$ 'at t x is F and has been F for exactly six months'. It follows from this definition that if something is F' at t it ceases to be F' immediately thereafter, simply in virtue of the 'passage' of t into the past. 'F'' seems clearly to be a McTaggartian predicate, like '10 years old', and one is inclined to say that something does not undergo genuine change in coming or ceasing to be F'.

---

[9] It may be objected that allowing for this sort of causality would complicate the scientific theories of these people so much that it would always be simpler for them to avoid the need of allowing for it by adopting an hypothesis according to which total freezes never occur. But this supposes that they *can* avoid the need for allowing it by adopting such a hypothesis. It seems entirely possible to me that they might find that in order to subsume even *local* freezes under causal laws they have to accept the existence of this sort of causality (e.g. they never succeed in explaining the termination of local freezes in terms of immediately preceding events in adjacent unfrozen regions), and that they might find other phenomena in their world that they are unable to explain except on the assumption that such causality exists.

But now suppose that the basic causal laws governing the world are such that the following is true: something's having been F for a period of one year is a causally sufficient condition of its becoming G at the end of that year (where 'G' is another non-McTaggartian predicate), and it is not the case that something's having been F for any interval of less than a year is a causally sufficient condition of its becoming G at the end of that interval. Given this, the causal implications of something's being F' at t are different from those of its being F at t; from the fact that something is F' at t, but not from the fact that it is F at t, we can infer that if it continues to be F for another six months it will then become G. And if we introduce another predicate 'F''', defining it like 'F'' except that 'exactly six months' is replaced by 'more than six months', we see that 'F'' and 'F''' are incompatible predicates having different causal implications. Now we are accustomed to regarding the causal properties of things, their 'powers', as intrinsic to them, and it is thus plausible to say that, when predicates differ as 'F'' and 'F''' do in their causal implications, then something does undergo genuine change in ceasing to exemplify one and coming to exemplify the other. But if we say this, then we will have to allow that, in remaining F for a year and not undergoing change with respect to any other non-McTaggartian property, a thing nevertheless undergoes genuine change. And this of course goes counter to the intuition that McTaggartian change is not genuine change. It remains true, I think, that the inhabitants of my imaginary world could have good reasons for thinking that there are intervals during which no non-McTaggartian changes occur—but given the sorts of causal laws they would have to accept in order for it to be reasonable for them to believe this, it is not so clear whether they would be justified, as I think we are in our world, in dismissing McTaggartian changes as not being genuine changes. The determination of whether this is so must wait on a closer examination of the considerations that underlie our intuitions as to the genuineness, or otherwise, of ostensible changes and properties.[10]

Supposing that it is possible for there to be time without change, how are we to answer the sceptical argument mentioned at the beginning of this paper—the argument that we can never be justified in believing that a given amount of time has elapsed since the occurrence of a certain event, since there is no way in which we can know that the interval between that event and the present does not contain one or more changeless intervals, perhaps lasting billions of years? I think the answer to this is that the logical poss-

---

[10] The need for this reservation was impressed on me by Ruth Barcan Marcus, who observed in a discussion of this paper that, if the 'mere' passage of time can itself have causal efficacy, it is not clear that it can be dismissed as not being genuine change.

ibility of such intervals, and the fact that such intervals would necessarily be unnoticed while they were occurring, do not prevent us from knowing that such intervals do not in fact occur. Given the nature of our experience of the world, the simplest theories and hypotheses that do justice to the observed facts are ones according to which changeless intervals do not occur. We do not indeed have a set of hypotheses that explain all observed phenomena, but none of the unexplained phenomena are such that there is any reason to think that positing changeless intervals would help to explain them. If our experience of the world were different in describable ways, e.g. if it were like that of the inhabitants of my imaginary world, then, so I have argued, it would be reasonable to believe in the existence of changeless intervals. But even then there would be no basis for scepticism about the measurement of time. The simplest set of hypotheses that did justice to the observed facts would then be one according to which changeless intervals occur only at specified intervals, or under certain specified conditions, where their existence and extent could be known (although not while they were occurring). If anything leads to scepticism it is not the claim that changeless intervals can occur but the claim that they might occur in such a way that their existence could never be detected. But it is not clear to me that even this is a *logical* impossibility, or at any rate that we must assert that it is in order to avoid scepticism. The claim that changeless intervals *do* occur in such a way that their existence cannot in any way be detected could not—and this is a logical 'could not'—constitute part of the theory that provides the simplest and most coherent explanation of the observed facts, and this seems to me a sufficient reason to reject it. It is 'senseless' in the sense that it could never be sensible to believe it; but it seems to me unnecessary to maintain, in order to avoid scepticism, that it is also senseless in the sense of being meaningless or self-contradictory. This is, in any case, irrelevant to what I have been arguing in this paper, for what I have suggested is that there are conceivable circumstances in which the existence of changeless intervals *could* be detected.

V

TIME, EVENTS, AND MODALITY

GRAEME FORBES

1

Newton betrayed none of Augustine's famous perplexity about the nature of time. He wrote: 'Absolute, true, and mathematical time, of itself, and from its own nature, flows equably without relation to anything external'.[1] But, according to Leibniz, the Newtonian or 'substantivalist' conception of time is subject to a serious objection, which he put as follows in his Third Paper in the correspondence with Clarke:

> Supposing any one should ask, why God did not create every thing a year sooner; and the same person should infer from thence, that God has done something, concerning which 'tis not possible there should be a reason, why he did it so, and not otherwise: the answer is, that his inference would be right, if time was anything distinct from things existing in time ... But then the same argument proves, that instants, consider'd without the things, are nothing at all; and that they consist only in the successive order of things: which order remaining the same, one of the two states, viz. that of a supposed anticipation, would not at all differ, nor could be discerned from, the other which now is.[2]

In this passage, Leibniz claims that Newton's substantivalism generates a distinction without a difference. The feature of substantivalism which Leibniz is exploiting is the idea that time is a container-like manifold in which any course of physical events *contingently* occupies the segment which it does. What guarantees the contingency is the Newtonian view that the time-series is an 'ontologically autonomous' substance, all of whose intrinsic and characteristic features are possessed independently of facts about the course

© 1993 Graeme Forbes
This essay has not been previously published in its present form. Parts of the essay are based on my 'Places as Possibilities of Location', *Noûs*, 21 (1987): 295–318. Other parts are extracted from a longer piece including discussion of Minkowski space-time based on work I did during a sabbatical leave from Tulane University spent at the Institute for Advanced Studies in the Humanities at Edinburgh University I thank Tulane for the leave which it granted me and the Institute for the ideal working conditions it provided.

[1] Scholium to Definition VIII of Isaac Newton's *Principia*, reproduced in *The Leibniz–Clarke Correspondence*, ed. H. G. Alexander (Manchester: Manchester University Press, 1956), 152
[2] *The Leibniz–Clarke Correspondence*, ed. Alexander, 26–7.

of physical events occupying a segment of the series. Ontological autonomy implies that for any possible world u there is another v exactly like u except that, in v, the course of history occupies a different segment of time from the segment it occupies in u; we say that v is 'temporally shifted' with respect to u. Leibniz then goes on to offer two reasons why the possibility of temporally shifted worlds is a *reductio* of any theory which accommodates it: (1) it would require God to have actualized the actual world without a 'sufficient reason' to prefer it over any of its temporally shifted counterparts; and (2) worlds which differ just by a temporal shift would be mutually indiscernible.

The family of views which relate time to temporal relations between events in some manner akin to that which Leibniz envisages is known as 'relationism'.[3] Leibniz's theological argument (1) for relationism has little appeal nowadays, but it can be recast in non-theological terms, as a condition of adequacy on explananda in science. According to this condition, any genuine fact can be differentially explained with respect to each of its parameters. Thus if an event occurs now, it must be explicable why *now*. But given the homogeneity of Newton's substantial time (each instant is indistinguishable from any other by its monadic properties and n-adic relations to times), there could be no explanation of why the course of events occupies this rather than that segment of it. So substantivalism generates explananda which fail the adequacy condition, and is for that reason to be rejected. However, I mention this secular version of (1) only to move on to (2), since the topic of explanatory adequacy would require a separate paper.

The objection (2) from indiscernibility retains its force.[4] In modern terminology, Leibniz is subscribing to a 'supervenience' principle: the facts about *when* events occur supervene on the facts about 'the successive order of things', which we may take to be the facts about temporal relations among events. Accordingly, two possible worlds cannot differ over the dates of events unless they also differ over temporal relations among the events. What non-theological justification could Leibniz's supervenience thesis possess? One might try to provide a justification in terms of some methodological principle, to the effect that apparatus which generates distinctions which cannot be discriminated by any empirical method is to be replaced by something less profligate. And the empiricist tradition in philosophy has long been concerned to formulate an acceptable version of such a methodological constraint.

---

[3] For a much wider-ranging discussion of relationism than is possible within the scope of this paper, see John Earman, *World Enough and Space-Time* (Cambridge, Mass.: MIT Press, 1989).
[4] Thus Earman: 'only dyed-in-the-wool absolutists would deny that Leibniz's argument has considerable intuitive appeal' (*World Enough and Space-Time*, 125).

However, there may be a better way to argue for the supervenience principle. For there are other supervenience principles which on *metaphysical* grounds reject the drawing of certain kinds of distinctions between possible worlds, distinctions which, on the face of it, are of a similar sort to the one Leibniz is rejecting here. Consider, for example, the hypothesis that there is a possible world *w* exactly like the actual world except that, in *w*, the Eiffel Tower does not exist. In its place there is another tower, the Schmeiffel Tower. But that is the *only* difference between *w* and the actual world: in *w*, the Schmeiffel Tower is made of exactly the same 18,000 pieces of steel as in the actual world, held together by exactly the same 2.5 million rivets. It is designed to the same specifications by the same Alexandre Gustave Eiffel, is erected at the same time and the same place in Paris, and in 1957 an extra 20.1 metres of construction are added to reach a total height of 321.75 metres, as in the actual world. As these respects of similarity between the Eiffel and Schmeiffel Towers are spelled out, it becomes harder and harder to continue to affirm the thought that the Eiffel and Schmeiffel Towers are different towers: the hypothesis that they are different acquires a certain degree of unintelligibility. But it is not very plausible that this unintelligibility traces to some methodological principle governing empirical enquiry. More likely, the hypothesis violates our concept of what it is to be a thing of the Eiffel Tower's kind. That the postulated difference seems unintelligible is a pointer to the fact that the identity of an artefact such as the Eiffel Tower is *exhausted* by such factors as its constitution and its design. To make sense of the idea that the actual world differs from *w* in the respect proposed, there would have to be more to the identity of the Eiffel Tower, something over and above all the features it shares with the Schmeiffel Tower (which are all its features period, except its identity). Call this extra element the 'primitive thisness' of the Eiffel Tower. If any genuine conception answered to the label 'primitive thisness', we could employ it to understand the putative numerical distinction between the Eiffel Tower and the Schmeiffel Tower. If we cannot make that distinction intelligible, this shows that the notion of primitive thisness is not applicable to artefacts like the Eiffel Tower.[5]

It is plausible that Leibniz's difficulty with the distinction between worlds which differ just by a temporal shift has a similar basis, whether or not Leibniz would have put it this way. If *u* and *v* are the same except that *u* has the Eiffel Tower where *v* has the Schmeiffel Tower, then there is an outer-

---

[5] For further discussion of the Eiffel Tower case and responses to objections, see my 'A New Riddle of Existence', forthcoming in James Tomberlin (ed.), *Philosophical Perspectives*, vii–viii: *Philosophy of Language and Logic* (Atascadero, Calif.: Ridgeview, 1993). The term 'primitive thisness' is from Robert Adams, 'Primitive Thisness and Primitive Identity', *Journal of Philosophy*, 76 (1979): 5–26.

domain isomorphism between u and v which is identity on the objects which make up other objects but non-trivially permutes the made-up objects. And if u and v are shifted worlds then there is an isomorphism between them which is identity on their events but permutes their times. In the former case, the isomorphism preserves the constitution–design relation in a way that is objectionable, while, in the latter, it preserves the occurrence relation in a way that is objectionable. But the metaphysical issue in the temporal case is rather less clear-cut. After all, a substantivalist about time could simply accept that there are shifted possible worlds, and deny that admitting the difference violates some aspect of what it is for something to be a time-series. For the entities on which the isomorphism is defined are in dispute in the artefact case (there is no such possible object as the Schmeiffel Tower), while not being in dispute in the temporal case. It is therefore up to the relationist to present reasons why Newton's account of time should be convicted of the same obscurantism as the doctrine of primitive thisness for artefacts; the example of the temporally shifted worlds, though it has some force, does not by itself establish the point.

Relationism also has a non-metaphysical motivation. For there is an *epistemological* problem with substantivalism: Newton's time-series does not exert any kind of causal influence, yet if the universe were Newtonian we would surely still have knowledge about time, for instance, about how much time elapses between two events. But it is a puzzle how we could acquire such knowledge, since it seems to be based on experience, while our mechanisms for acquiring knowledge based on experience are apparently causal. This puzzle echoes a similar one about our knowledge of mathematics, since, on the standard 'Platonist' view, numbers are not thought to exert causal influence, yet seem to be ontologically autonomous as well. However, it is likely that Platonism would lose its appeal if some other account of the nature of mathematics were forthcoming which was not beset by its own difficulties. Similarly, a detailed relationist theory of the dependency of time upon events would remove much of the appeal of substantivalism, which gets some of its attractiveness by default. We have seen that it is plausible that the identity of an artefact is determined by such features as its design and constitution. What we would like from the relationist is a theory of an analogous determination of the *time*-series by the *event*-series. Many different event-series should be able to give rise to the same time-series, of course, but the same event-series should not be able to give rise to different time-series.[6]

---

[6] It must be admitted that there is also an epistemological problem for relationism as it is to be portrayed in this paper. Heavy use will be made of modality, so that temporal knowledge will turn out to involve modal knowledge, which is itself something of a puzzle: how can we know what is

2

One way of proceeding is to portray the time-series as a logical construction
out of the event-series, in the same way that more complex number systems
can be logically constructed out of less complex ones.[7] For example, the
positive rationals can be constructed from the positive integers by identifying
rationals with certain set-theoretic constructs from integers and by showing
how the properties of rationals can be read off those of integers via the
construction. Each (positive) rational can be identified with an equivalence
class of pairs of integers; the definition is that for integers a, b, c, and d, the
pair $<a,b>$ is equivalent to the pair $<c,d>$ iff $ad = bc$ (juxtaposition is integer
multiplication). Arithmetical operations can be extended to rationals in terms
of this definition; e.g. for multiplication of rationals we say that if p and q
are rational numbers (i.e. equivalence classes of pairs of naturals), then
$pq = r$ iff, for any pairs of integers $<a,b>$ and $<c,d>$ such that $<a,b>$ belongs
to p and $<c,d>$ belongs to q, the pair $<ac,bd>$ belongs to r. Of course, there
is some sense in which a rational number is not *really* a set of pairs of
integers, but is rather some kind of abstraction from it. The nature of this
relationship is not easy to explain, but, whatever it is, the relationist can say
that the same relationship holds between an instant and the equivalence class
with which the instant is identified on her construction.

   The idea is to regard the time-series as an abstraction from the event-series.
Provided the abstraction process is fully deterministic, we could not obtain
different time-series from the same event-series, and so we exclude temporally shifted worlds. As a first attempt at working this out, we will identify
moments of time with equivalence classes of events in the following way.
The construction is relativized to possible worlds, in the sense that each
world's time-series is constructed from within, the ontology of the construction at a world being the events which occur at that world. We assume these
events to be instant events, since it introduces irrelevant complexity to use
events with non-zero duration. Secondly, the relational basis of the construc-

possible, when our senses only put us in touch with what is actual? However, there are two reasons
why substituting the problem of modal knowledge for the problem of knowledge about causally
isolated entities is an advance. First, there is going to be a problem of modal knowledge anyway,
so two problems have been reduced to one. Secondly, we have some idea of how to make progress
with the problem of modal knowledge in a reasonable way—see Stephen Yablo, 'The Real Distinction between Mind and Body', *Canadian Journal of Philosophy*, suppl. vol. 16 (1990): 149–201—
while accounts of knowledge of causally isolated entities which keep them causally isolated
inevitably postulate a mysterious special faculty that is independently unmotivated.

   [7] For an account of the successive construction of the integers, the rationals, and the reals, see
Robert Stoll, *Set Theory and Logic* (New York: Dover, 1961).

tion will be (1) the collection of order facts about events, which should determine a weak linear order on the set of events (for any two events, either one is before the other or they are simultaneous); and (2) all the comparative facts about the magnitudes of elapsed intervals between events, facts such as 'twice as much time elapsed between lightning flash f_2 and thunderclap h_2 as between lightning flash f_1 and thunderclap h_1'. The totality of these facts suffices, relative to a choice of unit, to assign scalar magnitudes to the intervals between events. Note that this basis for the construction is epistemologically unproblematic, since a normal human being is capable, just on the basis of experience of the world, of knowing order facts and (at least approximately) comparative-magnitude facts. The first proposal, therefore, is to take instants of time to be simultaneity classes of events, to read the order of the instants off the order of the events, and the metric on the time-series off the metric on the event-series.

This abstraction certainly eliminates temporally shifted worlds. For if the courses of events at u and v are the same, then so are the simultaneity classes, and if these just *are* the instants of time, or even if they are merely related to them as a rational number is related to its equivalence class, then any event in one of the worlds occurs at the same time as it does in the other. But this construction of times has an obvious flaw: it will make it impossible for different things to happen in two worlds at the same time. Suppose, for example, that in u a bomb goes off at p at t while in v it fails to go off at p at t. Then in u and v the t-simultaneity classes are different classes, since one contains an explosion that the other does not.[8] So it is not possible that, at time t, *that* explosion does not occur; but this is surely absurd.

Any relationist construction will run into some version of this difficulty if it restricts itself, in constructing the time-series of a world w, to goings on in w. This restriction will also cause problems if one wishes to allow for intervals of empty time between events or for a time-series which outstrips the event-series, since only one instant of eventless time is available, corresponding to the empty set of events.[9] To meet the difficulty, the basis of the construction of the time-series of a world has to be expanded to allow facts about goings-on in other worlds to play a role. Yet we do not want the events of other worlds to be in any sense *constituents* of the times of a given world,

<hr />

[8] In this argument I am making some relatively uncontroversial assumptions about transworld identity for events

[9] For discussion of the coherence of the idea of empty time between events, see Sydney Shoemaker, 'Time without Change', *Journal of Philosophy*, 66 (1969): 363–81 [Essay IV in this volume], and W H. Newton-Smith, *The Structure of Time* (London: Routledge & Kegan Paul, 1980), ch. 2. Granted that an anti-Humean view about causality could allow for stretches of eventless time as a consequence of causal properties possessed by events, relationism should be neutral on whether or not empty time between events is possible.

since the events of other worlds may not occur in the given world. It would be problematic to identify something which exists at a world with something that has constituents which do not exist at that world.[10]

These desiderata can be met if we restrict our attempts to provide a common time-series for possible worlds to worlds which fall into the same 'branching equivalence class'. Think of a world as consisting in a sequence of events, that is, a course of history. Then two worlds may share an initial segment of their courses of history, diverging from each other only after a certain point. Such worlds are called branching worlds. The relation 'branches from' is reflexive by stipulation (the irreflexive relation is 'proper branching'), trivially symmetric, and is easily seen to be transitive, so the relation partitions possible worlds into groups in a mutually exclusive and jointly exhaustive way. Then we can construct the time-series of a world w as a sequence of equivalence classes of entities which exist at w, but using the facts about the worlds which branch from w to determine the classes.

The revised construction for a world w expands the basis of the construction to include the order facts about events, and the comparative facts about the magnitudes of elapsed intervals between events, in all metaphysically possible worlds branching from w. We can assign scalar magnitudes to intervals between events in w much as before, using a unit that is common to all the worlds of w's branching class. For if e_1 and e_2 are events in the common segment of u and v whose separation is used as a unit, then if w' branches from u and v before e_1 and e_2 have both occurred, we can still apply the unit $d(e_1, e_2)$ (the magnitude of the interval between e_1 and e_2) to w', by comparing the relative magnitude of $d(e_1, e_2)$ with the separation of some events e_3 and e_4 which both occur in w' before the branch from u and v. Next, for the times of w we take equivalence classes of ordered pairs $<e, m>$ consisting in an event e and a magnitude m: there is an equivalence class $[<e, m>]$ in the time-series of w iff there is a world u which branches from w, e belongs to the common initial segment of u and w, and in u there is some event which occurs m units of time after e. The equivalence classes themselves are generated by the relation defined by: for any e_1 and e_2 which occur at w, $<e_1, m_1>$ is equivalent to $<e_2, m_2>$ iff either e_1 precedes e_2 and $m_1 = d(e_1, e_2) + m_2$, or e_2 precedes e_1 and $m_2 = d(e_1, e_2) + m_1$. This is easily seen to be a genuine equivalence relation, and we determine a strict linear order for the classes by the definition: $[<e_1, m_1>]$ precedes $[<e_2, m_2>]$ iff for some e, $<e, m> \in [<e_1, m_1>]$ and $<e, n> \in [<e_2, m_2>]$ and $m < n$. Again, it is easily seen that the resulting order does not depend on choice of e. Properties

---

[10] For some of the general issues here, see Kit Fine, 'First-Order Modal Theories I: Sets', *Noûs*, 15 (1981): 177–205.

of time, such as the various aspects of homogeneity, are consequences of the construction. For instance, if the time-series of a world w is continuous, this is because, for any event e in w and any real magnitude m, there is a world u branching from w after e occurs and in u there is some event e' later than e such that $d(e, e') = m$.

This provides each world w in a branching class with its own time-series, and allows for empty time after events have run their course in w, or even between events in w, in virtue of the reference to the events of branching worlds where, intuitively, events occur at those times which are empty in w. The construction, we might say, portrays instants as possibilities of occurrence. And, despite its use of possible worlds, the construction is actualist: worlds do not enter into its ontology, and its ideology can be expressed in a modal language with actualist quantifiers and a predicate 'W' for being a world which is satisfied at each world w by w alone. However, there is a question about the membership of a branching class. Our procedure relative to a world w is to use the temporal facts about all metaphysically possible worlds branching from w, but intuitively, if these worlds include worlds with a different type of time-series, the results will not be consistent. For instance, w may be a Newtonian world with a time-series that has an intrinsic metric, while u branches from w but is a 'Machian' world in which time has no intrinsic metric.[11] One response to this is that branching brings with it a guarantee that worlds in the same branching class have the same *type* of time-series. This is so because it could not literally be the *same* subsequence of events that the two worlds share if time is structurally different in them. (On this view, if branching is also necessary for having objects in common,[12] the properties of time at a world are essential to the players in the history of that world: if the universe had been Machian, we would not have existed.) Alternatively, we can regard branching as a subrelation of an accessibility relation, T-accessibility, which holds between worlds with the same type of time. For the substantivalist, if u and v are T-accessible, this is *because* they have time-series with the same intrinsic properties, whereas, for the relationist, the latter fact is a *consequence* of T-accessibility, which is either primitive or explained in other irreducibly modal terms. This divergence over T-accessibility echoes other disputes between those who posit entities of a certain sort and those who would construct them. For example, according to some, if 'S and S' mean the same' is true, that is because there are meanings π and π' such that S has π and S' has π' and $\pi = \pi'$; while, according to others,

---

[11] See Earman, *World Enough and Space-Time*, 28–30.
[12] I argue that it is necessary in my *The Metaphysics of Modality* (Oxford: Oxford University Press, 1985), 148–52

there is a more fundamental 'samesaying' relationship in terms of which talk about meanings is to be explained.[13]

There is still the difficulty of providing for the intuition that most events might not have occurred at the instants at which they in fact occurred, i.e. the intuition that the date of an event is contingent, at least in the sense that the event might not have occurred at all (we need not take a position on whether, *if* an event occurs at two worlds, it must occur at the same time in both, i.e. on whether the time of an event is essential to it[14]). To provide for the contingency of the state of the world at a time, we need some notion of 'transworld identity' for times. But the equivalence classes of two branching worlds u and v which, intuitively, represent the same time, need not be literally the same equivalence classes, since either may contain an element $<e, m>$ which the other does not, e not being a constituent of the other's event-series. However, literal transworld identity of equivalence classes is more than is required to save, for example, the intuitive thought that *this* explosion might not have occurred now. This thought is expressed in modal language, the modal verb 'might not have occurred' containing the modal operator $\Diamond$. It is true that the standard possible-worlds formulation of the thought is that there is a world w in which, at *this* time, the explosion does not occur; which, understood in the intended way, requires the present instant *itself* to be a member of the time-series of w. But instead of a possible worlds framework in which the referring terms of the modal language are interpreted as standing for the very same things at different worlds, we can interpret them in a framework which is at least partially *counterpart-theoretic*. In such a framework, a temporal term is taken to refer at worlds u and v to things which are counterparts or 'representatives' of each other, and representatives need not be numerically identical (identity is a special case of representation).[15] Thus our intuitive thought about the explosion, that it might not have occurred now, really only requires for its truth a world where the explosion does not occur at the instant of time in that world's time-series which is the *counterpart* of the instant of time at which it actually occurred. And it is easy to define the appropriate counterpart relation for branching worlds u and v: a time $[<e_1, m_1>]$ of u is the counterpart of time $[<e_2, m_2>]$ of v iff, for some e in the common initial segment of u and v, there is m such that $<e,m>$ belongs to both $[<e_1, m_1>]$ and $[<e_2, m_2>]$; this just means that as much time elapses in u between e and e_1 as elapses in v between e and e_2.

---

[13] See Donald Davidson, 'On Saying That', in Donald Davidson, *Inquiries into Truth and Interpretation* (Oxford: Oxford University Press, 1984), ch. 7.

[14] The whole question of the essences of events is rather perplexing. See Jonathan Bennett, *Events and Their Names* (Oxford: Clarendon Press, 1988), ch. 4.

[15] Counterpart theory is due to David Lewis. See his *On the Plurality of Worlds* (Oxford: Blackwell, 1986), 1–13, for an accessible account.

An attractive feature of the proposal is that it provides a clear answer to the question in virtue of *what* does such and such a world represent the state of affairs that the explosion does not occur *now*.[16] A requirement that there should be something in virtue of which it is this temporal state of affairs rather than that one which is represented by a world appears to underlie Leibniz's rejection of shifted worlds. That is, the implicit supervenience principle (no difference in dates without some difference in temporal relations) is motivated by the thought that there is nothing which could *make it the case* that a given non-actual world represents the actual world's history as shifted a year, since nothing could make it the case that the counterpart relation holds between times across two shifted worlds if these times are not the times of the same events. A substantivalist might query the legitimacy, or even the intelligibility, of the demand for something which makes it the case. But suppose A meets B, an old friend whom A has not seen for many years, in the street. Then we can intelligibly ask what makes it the case that the person whom A now meets is the same person as the one she roomed with in college. Is it, for instance, that B now has *the same hairstyle* as her room-mate had then? One who thinks this is clearly the wrong answer has at least understood the question, establishing its intelligibility. And to hold that *no* conditions are sufficient, even if some are necessary, is to impute to personal identity a primitive thisness, comparable to the we-know-not-what which distinguishes the Eiffel and the Schmeiffel Towers. There is surely some merit in pursuing theories which are free of such occult elements. Thus both substantivalist and relationist should prefer to have an account of what makes it the case that a world verifies that a particular condition applies to *this* time or time interval rather than *that* one. The answer is straightforward for the substantivalist, more complex for the relationist, but the construction we have sketched seems to satisfy the desideratum.

How does this relationist construction rule out shifted worlds? One objection which it suggests is that there simply cannot be worlds that have the same course of events but are nevertheless distinct worlds. If the worlds have the same course of events and their spatio-temporal features are logical constructions, in what could their difference consist at all? However, this objection seems to take sides on the issue of Humean theories of causal powers (see also note 9): perhaps, despite the sameness of the course of events, the structure of causality is not the same in the two worlds. A more intrinsically relationist objection to shifted worlds is this. Such worlds by

---

[16] In my view, a criterion of counterparthood in terms of overall similarity, or even in terms of similarity in theoretically important respects, does not provide for the correct range of possible worlds, since we want to admit worlds where the counterpart relation associates times in a way that does not respect overall or theoretically important respects of similarity.

hypothesis do not have any initial segment of their histories in common, since everything is supposed to occur in one at a different time from when it occurs in the other. Consequently, if there are two shifted worlds, then they do not belong to the same branching class. But this means that sameness or difference in time of occurrence is not even *defined* for them. Therefore, there can be no fact of the matter as to whether or not the event-series of one world occupies the same or a different interval of the time-series as in the other, since there is no intelligible notion of a common time for the two worlds.

3

There are a number of difficulties for our relationist construction which focus on the branching requirement. First, it could be objected that the requirement is too restrictive, since it rules out (1) empty time before the course of events begins, and (2) possible worlds in which time passes but no events ever occur. But to accommodate (1) would be to reintroduce shifted worlds, since, if a world can have a stretch of empty time followed by a course of events, it is hard to see why another world could not have a longer or shorter stretch of empty time followed by the same course of events. (2) is also something failing to provide for which need not embarrass us. For it is unclear exactly what the possibility amounts to, and it seems likely that the conviction that it is a possibility is tied essentially to the substantivalist conception of time as an ontologically autonomous container. Perhaps there are space-time theories with empty models, but such models, which Earman suggests may not be physically possible,[17] can be regarded as artefacts of the formalism.

Secondly, our definitions require that the *same* unit of time be used to assign magnitudes to the intervals between events in any two branching worlds. Therefore, they do not support comparison of the times of two worlds u and v which have only a *first* event in common, as opposed to an initial segment of events, for with only a first event in common between u and v there is no common interval to use in assigning commensurable scalar magnitudes to all event intervals in u and v, and so no way of defining counterparthood for the instants of u and v. But this is an objection only if there *should* be a common time definable for two such worlds. If there is no real need to provide a common time in this situation, we can simply define 'branching' so that two worlds with at most a first event in common are not counted as branching worlds.

---

[17] Earman, *World Enough and Space-Time*, 115

This brings us to the more general question of whether the branching requirement for assigning commensurable scalar magnitudes is too restrictive. For it certainly seems that we can consider a possible world totally different from the actual world, one which, say, conforms to a cosmological model which is actually false (e.g. a steady-state model, if it is false), and in which events unfold for more than a year. It will not do for the relationist to allow such a world on the grounds that many worlds can conform to a cosmological model and there is sure to be at least one in which the course of events exceeds a year's duration. The existence of a common interval is supposed to be that which makes it the case that intervals in a non-actual world have such and such a magnitude, specified in units defined by intervals between actual events: the ratio relationships between intervals in the common initial segment of two worlds and intervals after they have branched underpins the application of a common unit applicable across the worlds throughout their respective courses of events. So for each world which does not branch from the actual world, there is nothing which makes it the case that an interval between events in it is x years long rather than y years long. Thus, apparently counter-intuitively, there is no non-branching world of which it is determinately true that the course of events exceeds a year's duration.

However, it is possible to qualify the relationist view to allow a non-branching world to make true propositions about the magnitudes of intervals specified in units defined by actual world intervals. For we can distinguish the making true of propositions about the magnitudes of intervals between events from the making true of propositions about *specific* intervals (or instants). The hypothesis that some actual moments of time t_1 and t_2 could have occurred in the reverse order is an example of a *de re* modal hypothesis such that, concerning any purportedly verifying world, one would certainly want to press the question what makes it the case that the purportedly verifying instants really are the counterparts of the actual instants. A branching model should be insisted on since the branching requirement has clearest applicability to worlds which are supposed to have entities in common, including both ordinary objects and instants of time. But two worlds having intervals measurable in the same units is weaker than their having times in common; so perhaps we can permit a pure stipulation about the duration in years of an interval between events in a world which is wholly alien to the actual world, in the sense that the intersection of its basic ontology with the actual world's is empty. More generally, we can concede the intelligibility of metrical stipulations for worlds which are mutually wholly alien (so no two have a common initial segment). This concession does not affect our rejection of shifted worlds, though of course it would not satisfy someone who holds that we can make coherent modal hypotheses about what could be happening *this* year, hypotheses which, because of their content, require mutually alien worlds to make them true. Fortunately, excluding this seems much less counter-intuitive.

4

The construction is also open to various circularity objections. Two related objections are that the modal facts to which the construction appeals are grounded in actual facts, and that the actual facts to which the construction appeals are grounded in facts about substantival time. If it is true that there *could* be an event in such and such temporal relations to actual events, it will be said, that must be because of actual facts, and the only actual facts which suffice are facts about the actual time-series. Similarly, if it is true in the actual world that twice as much time elapsed between lightning flash f_2 and thunderclap h_2 as between lightning flash f_1 and thunderclap h_1 that, it will be said, must be in virtue of f_1, h_1, f_2, and h_2 occurring at times t_1, t_2, t_3, and t_4 such that $d(t_3,t_4) = 2d(t_1,t_2)$. This is particularly so if part of the reason why $d(f_2,h_2) = 2d(f_1,h_1)$ is that there is a stretch of eventless time between f_2 and h_2. However, these objections simply fail to take the relationist project seriously. Though some philosophers hold that all modal facts must ultimately reduce to actual facts, this idea has not been taken beyond the stage of slogan. It is hardly intrinsically perverse to begin with modal facts and proceed from these to a new level of actual facts. Similarly, there is nothing intrinsically perverse about beginning with comparative-magnitude facts and using these to generate instants and the intervals between them, rather than the other way round.

A more interesting group of circularity objections concerns the nature of the events and event-sequences which the construction employs. If the construction is faithful to the nature of time, then this means that events are ontologically prior to instants. Yet that cannot be, it may be objected, since instants are required for the (intraworld) individuation of events.[18] For example, consider some regularly repeated phenomenon, such as the earth's orbiting the sun, and suppose that each orbit traces the same path through space. Then what distinguishes one orbit from another is the fact that they occur at different times. However, the relationist can reply to this that, in place of times, events may be individuated (at least in part) by their temporal

---

[18] For a persuasive account of intraworld event-identity which uses times, see Lawrence Brian Lombard, *Events* (London: Routledge & Kegan Paul, 1986), esp. ch. 6, sects. 1–2.

relations to one another, or, following Davidson,[19] by their causal relations to one another: the proposal would be that if e_1 and e_2 are the same event, this is in part because each stands in the same temporal relations to any given event as the other, or because each stands in the same cause–effect relations to any given event as the other. Are such accounts of event individuation illegitimate because one of their conditions quantifies over events, items of the kind being individuated? This makes the account epistemologically unhelpful, since someone unsure of how to individuate events could not in general apply the account to resolve problem cases: each application would potentially raise a further difficulty of the same sort. But as accounts of what event-identity *consists in*, neither proposal can be faulted on the grounds of its form. The idea behind both is that there is a network of relations of a certain sort, and events just are the things which sit at the nodes of the network. There would be a problem only if identity conditions for particular events were presupposed by an account of what a causal or temporal relation is.

The most interesting circularity objections have been saved for last; they concern how a standard problem for relationism, which Teller has labelled the 'problem of inertial effects', arises within the present framework.[20] We have spoken of possible worlds as 'courses of events' and have proposed to construct a classical time-series for a world w from temporal relations amongst events in the worlds branching from w. But what if the correct individuation of *courses* of events requires that we have available properties and relations into which time enters? Suppose w is a Newtonian world consisting in two parallel rings r_1 and r_2 of material objects, relatively at rest as some force maintains them in this configuration. Then the rings begin to rotate relative to each other, yielding a sequence of simultaneity classes of events in a certain temporal order, each event in a class being an event of one object in one ring assuming a particular spatial relation to another in the other ring. In Newtonian mechanics, given one total temporally ordered sequence of such classes of events, there are non-denumerably many genuinely distinct possible worlds realizing it. For instance, there is the world u, where r_1 does not rotate at all while r_2 rotates anticlockwise, and there is also the world v, where r_2 does not rotate at all while r_1 rotates clockwise, the angular velocities of the rotating rings in the two cases being the same.

[19] See Donald Davidson, 'The Logical Form of Action Sentences', in Donald Davidson, *Essays on Actions and Events* (Oxford: Oxford University Press, 1980). For a reliable short guide to many of the issues surrounding the intraworld individuation of events, see Terence Parsons, *Events in the Semantics of English* (Cambridge, Mass.: MIT Press, 1990), ch. 8.

[20] See Paul Teller, 'Substance, Relations, and Arguments about the Nature of Space-Time', *Philosophical Review*, 100 (1991): 363–97.

w's branching class has to include u and v.[21] But how is the relationist to conceive of the difference between worlds where the relative rates of rotation between the two rings are the same? He cannot just ascribe different accelerations to the objects in the rings in different worlds. Though that indeed is how the worlds differ—for instance, the objects in the outer ring really are accelerating in u and really are not in v—acceleration is the measure of rate of change of velocity with respect to time, and so acceleration facts cannot be used in the construction of time.

But there is no commitment in the kind of constructionist relationism which we have been developing to the thesis that all motions must be manifested by some characteristic pattern of changes in spatial relations through time. If u and v differ as imagined, then there will accompanying differences which betray the motion differences: the pattern of forces acting in u must be quite different from the pattern of forces acting in v, and the relationist can include the impressings of such forces among the events belonging to the courses of events which constitute u and v. However, this is a satisfactory solution to the general problem only if, whenever two branching worlds differ in ways which can be characterized in terms which presuppose time (or, for other relationist projects, space or space-time), there is an alternative characterization in terms without such presuppositions, for instance in terms of the actions of forces. Fortunately, there is reason to believe that this condition is met. For if the condition were to fail, this means that there would be frames of reference which are equivalent for the purposes of science but within which certain motion facts are in principle undetectable, in the very strong sense that keeping the space-time fixed, no *metaphysically* possible experiment would detect them. However, following Earman, we can impose an adequacy constraint on a theory of motion, i.e. on an account of what the physically possible courses of events are, which requires that physical laws and space-time structure be 'appropriate' for one another.[22] This constraint

---

[21] Strictly speaking, if w is Newtonian, u and v cannot be, since Newtonian universes are deterministic, so any course of events that starts out like w's must continue to be like w's. This is a general feature of the branching conception: one has to think of all but one of the worlds in a branching class as containing a momentary violation of certain laws (though of course not *its* laws). Therefore, u and v are to be thought of as being as Newtonian as possible.

[22] A *space-time symmetry* of a theory T is a permutation of the manifold of a model of T which preserves differential structure and under which the spatio-temporal properties and relations of the model are sent to themselves, while a *dynamic* symmetry is a permutation which is identity on space-time properties and relations and under which the dynamic properties and relations are mapped to properties and relations which are also consistent with T. Earman's proposal, as I understand it, is that a theory of motion T and a space-time manifold-type M are appropriate for each other iff M's space-time symmetries are exactly T's dynamic symmetries (*World Enough and Space-Time*, 45–6). Thus, for example, Newtonian space-time and Newtonian mechanics are inappropriate for each other, since there are dynamic symmetries of the mechanics which are not symmetries of the space-time (the spatial distance between events at different times is intrinsic to Newtonian space-time but varies from inertial frame to inertial frame).

rules out differences between at least some courses of events that are describable only with reference to a time or space-time manifold, i.e. which have no dynamic manifestation. What remain untouched are larger departures from physical into metaphysical possibility. In the previous paragraph, u and v were understood as being as Newtonian as possible (see note 21), but are there not also worlds branching from w where the rotations are different but the differences have no dynamic manifestation? Perhaps such a thing is metaphysically possible, or at least conceivable: u branches from w as one ring begins to rotate, for no reason at all, while v branches as the other ring begins to rotate, for no reason at all (or at least no reason that is differential as regards u). If so, the relationist must parley the existence of such worlds into a further constraint on which branching worlds figure in the construction of a given world's time.

There remains the threat of *explanatory* circularity. For example, some aspects of a course of events in a manifold may be best explained by positing a field which is defined for every point of the manifold. For the constructionist relationist, fields are not themselves obviously problematic, since prima facie a constructed manifold can support a field as well as an ontologically autonomous one. Nevertheless, there may be a sense that the direction of explanation is at odds with the direction of construction. Or it may be that some structural feature of the relevant manifold itself directly enters into the explanation of aspects of a given course of events with it.[23] Are ontologically autonomous manifolds required for explanations of this sort to be genuine?

If genuine explanation is just a matter of giving an economical systematization which satisfies conditions of empirical confirmability and disconfirmability, then ontological autonomy would presumably not be a well-motivated requirement. On the other hand, perhaps more is required of an explanation if it is to be genuine—the nature of the whole story here, which no one knows, is a central issue in the philosophy of science. For relationism to be compatible with any reasonable outcome, we would have to be able to construe explanatory appeal to a manifold, or to entities whose characterization is given via a manifold, as shorthand for something more elaborate involving the modal facts from which the manifold in question is constructed. If it seems odd that such facts play a role in genuine explanation, the relationist can at least venture to suggest that it is no odder than giving the role to an ontologically autonomous manifold.

---

[23] See Teller, 'Substance, Relations, and Arguments about the Nature of Space-Time', 376–9.

3

THE DIRECTION OF TIME

VI

UP AND DOWN, LEFT AND RIGHT, PAST AND FUTURE

LAWRENCE SKLAR

1

Few philosophical theses match the dramatic impact and striking illumination of Boltzmann's brilliant speculation about the reducibility of the intuitive notion of the direction of time to features of the world characterizable in terms of the theory of order and disorder summarized in the notion of entropy. Taking the progression of isolated systems toward states of highest entropy, characterized phenomenologically by the second law of thermodynamics, and given a far deeper explanation in his own theory of the statistical mechanics of irreversible processes, Boltzmann suggested that rather than viewing these theories as merely describing the asymmetric change of the world from past to future, we should find in them the very basis of our concept of the distinction between the past and future directions of time.

Building on Boltzmann's rather sketchy remarks, Reichenbach, in what many consider to be his most distinguished contribution to the philosophy of physics, elaborated for us a highly complex and subtle account of the entropic theory of time order. Yet despite Reichenbach's very impressive efforts, and the further illuminating work of others who have followed him such as Grünbaum, Watanabe, Costa de Beauregard, and others, the claim that the very notion of temporal asymmetry reduces to that of the asymmetry of entropic processes in time remains, to say the least, controversial. To some it seems obviously true in broad outline, whatever details still need filling in. To many others the very idea of the programme seems prima facie absurd.

While much remains to be done on the 'physical' side of this issue, in the way of providing for us a single coherent physical account of the source of

Lawrence Sklar, 'Up and Down, Left and Right, Past and Future', first published in *Noûs*, 15 (1981): 111–29. Work on this paper was supported in part by a research grant from the National Science Foundation, grant no. SOC 76–22334. The text as reprinted here includes some small changes made by the author when this essay was reprinted in his *Philosophy and Spacetime Physics* (Berkeley, Calif.: University of California Press, 1985). Reprinted by permission of the author.

entropic asymmetry, and in the way of definitively characterizing the physical connection of this asymmetry with the other fundamental temporal asymmetries of the world such as the outward radiation condition and the cosmic expansion, I believe that some insight into the roots of the persistent controversy can also be obtained by an attempt to become a little clearer concerning some philosophical aspects of the question about which we are yet not as clear as we might be. In particular, I think we need to be far clearer than we have been on the question of in just what sense the entropic theory is claiming that the very meaning of assertions about the direction of time are 'reducible' to assertions about entropic processes. At least two fundamentally different notions of meaning reduction are available to us, and I think that confusion about just which sense the entropic theorist has in mind has served to cloud the issues in a significant way.

As a means of access into this problem I would like to make some comparisons between three different asymmetries in the world: that between the upward and downward directions of space, that between left- and right-oriented systems, and that between the past and future directions of time. I think that exploring the analogies and disanalogies between these three cases of 'asymmetry' may make it clearer to us just what the entropic theorist is really claiming. While I don't believe that the insights we gain will resolve the question as to whether or not the entropic theorist is *right*, perhaps we will be a little clearer about just what both he and his opponent have a right to claim as evidence for and against the reductionist position.

2

We have a concept of the left–right distinction; and more, individual concepts of left- and right-oriented systems. We can properly identify left- (right-) handed objects; train others to do so; communicate meaningfully using the terms ('Bring me the left-handed golf-clubs'); etc.

Now there may be physical phenomena described by laws which are not left–right symmetric. Current physical theory postulates that this is so, as is revealed in the familiar examples of the parity non-conservation of weak interactions; although whether this will remain the case at the level of the 'most fundamental' laws remains an open question. Certainly there are many phenomena in the world which are non-mirror-image symmetric in a *de facto* rather than lawlike way, e.g. the preponderance of dextrose over levulose, etc. But do any of these asymmetries of the world in orientation have anything to do with what we *mean* by left and right? Is there any plausible sense in which the orientation concepts 'reduce' to concepts of a not prima facie spatial orientation sort?

Most of us think not. Of course it is the case that if we wish to teach the meaning of, say, 'left' to someone, without transporting to him a particular left-oriented object, we would need to do so by means of one of the familiar features of the world lawlike or *de facto* associated with orientation ('Left is the orientation in which . . .'). Even then, as we know, there are the difficulties which reside in assuming that the association where he is is the same as that where we are. (What if he lives in an anti-matter world and CP invariance holds? What if there is more levulose than dextrose on his planet?) And, of course, if one is taking the (dubious) line that a mirror-image possible world would be the same possible world as the actual one, one would have to assume that in this mirror-image world the mirror laws and *de facto* correlations hold to make the Leibnizian argument (qualitative similarity implying possible world identity) go through. But none of this is sufficient to back up in any way the claim that left-handedness just is, or that 'left' just means, some relation (term) expressible in terms of not prima facie orientation concepts.

Suppose, for example, that some fairly substantial miracles occurred in this world. All of a sudden electron emission from spinning nuclei begins occurring with the dominant emission in the opposite direction from the present preferential axial direction. Would we then say that the clockwise direction had become the counter-clockwise? That right-handed gloves had suddenly become left-handed? Nothing of the sort. We would indeed be astonished and look desperately for some explanation of this mirror reversal of a law. But we would, I believe, still take it that we could recognize left-and right-handed objects as before, teach the meanings of orientation terms by ostension as before, etc.

We believe that orientation is just a basic geometric property of an oriented system. It is epistemically available to us in as direct a manner as is any geometric feature of the world. The meanings of our orientation terms are fixed for us by ostension of particular oriented objects and our facility for abstracting the right property intended by the teacher. Nothing about weak interactions is relevant here in any way. It is merely an empirical discovery of a lawlike correlation that takes place when we discover that as a matter of fact weak interactions take place in an orientation-discriminating way. Were we to live in a universe in which as a matter of law, or merely as a matter of pervasive facts, all red objects were square and all square objects red, this alone would hardly be grounds for saying that redness was squareness nor that 'red' meant 'square'.

Of course our notions of handedness are complicated by reflections on the facts about dimensionality and global orientability which have been frequently pointed out. Prior to realizing the possibility that space might have a fourth spatial dimension and might be globally non-orientable we fail to notice the distinction between the partition of handed objects into two disjoint classes relative to their being constrained to the subspace in which they are embedded and the local region of that subspace, and the global distinction which would be well defined only if the subspace exhausts the full dimensionality of space and only if that space is orientable. Under the impact of this new awareness we may want to distinguish full handedness from what I have previously called local three-handedness.[1] And we may wish to say that the intuitive concept we had all along was the latter rather than the former. But none of this additional complication, I believe, vitiates the point here that the facts about what non-orientability features of the world are lawlike or *de facto* associated with handedness are irrelevant to the conceptual analysis of what we meant all along by handedness attributions.

3

Considerations like those above might lead us by analogy to make the parallel remarks about entropy increase and the future direction of time. Isn't the case just like that of weak interactions: we discover a pervasive correlation in the world, this time one whose status while not lawlike is not easily thought of as merely *de facto* either? But why should this in any way lead us to think that the very concept of futurity reduces in any sense to that of entropic increase?

But Wittgenstein has warned us against the deficiency diseases caused by an unbalanced diet of analogies, and we would be well advised before making a hasty judgement to look at another case which may be viewed as providing an analogy supportive of just the opposite view about time and entropy.

The suitable dietary supplement is provided for us in Boltzmann's elegant if sketchy presentation of his position in the *Lectures on Gas Theory*:

One can think of the world as a mechanical system of an enormously large number of constituents, and of an immensely long period of time, so that the dimensions of that part containing our own 'fixed stars' are minute compared to the extension of the universe; and times that we call eons are likewise minute compared to such a period. Then in the universe, which is in thermal equilibrium throughout and therefore dead, there will occur here and there relatively small regions of the same size as our galaxy (we call them single worlds) which, during the relatively short time of eons, fluctuate noticeably from thermal equilibrium, and indeed the state probability in such cases will be equally likely to increase or decrease. For the universe, the two directions of time

---

[1] See Lawrence Sklar, 'Incongruous Counterparts, Intrinsic Features and the Substantiviality of Space', *Journal of Philosophy*, 71 (1974): 277–90.

are indistinguishable, *just as in space there is no up and down. However, just as at a particular place on the earth's surface we call 'down' the direction toward the center of the earth, so will a living being in a particular time interval of such a single world distinguish the direction of time toward the less probable state from the opposite direction (the former toward the past, the latter toward the future).*[2]

Past and future, then, are to be viewed like up and down, and the progression toward higher entropy states like the direction of the gradient of the gravitational field (the obvious generalization of the direction of the centre of the earth). It is well worth our time then to ask what we do and should say about the relationship of up and down to directions characterized in terms of gravitation and to ask for the grounds of the position we do take. We must then ask whether things are just that way with past and future and temporal directions picked out by entropic features of the world.

4

We have a pre-scientific, pre-philosophical understanding of the distinction between the upward and the downward direction of space. We can communicate with these concepts since they are teachable and suitable for an inter-subjective language. We could teach the concept to someone either by an ostension which relies on observation of the behaviour of objects (by and large they move, when unsupported, in the downward direction) or by reliance on our internal 'sense' of the downward direction, using our sense of this direction to pick it out and then ostending it to another who can then identify it again by his own internal 'sense' of down. Naïvely we view it as a global notion, in the sense that parallel transport of a downward-pointing vector keeps it downward-pointing.

But then we discover gravitation. We come to understand that it is the local direction of the gradient of the gravitational field (on the surface of the earth, the local direction of the centre of the earth) which 'picks out' at any point the downward direction. 'Picks out', though, in a deep sense. It isn't just that the local gradient happens to point down, nor even that the local direction of the gravitational gradient points downward as a matter of lawlike necessity. Rather it is the reference to the local behaviour of the gradient of gravity which offers a full and complete account of all those phenomena which we initially used to determine what we meant by the downward direction in the first place.

Understanding gravity we understand why, in general, objects fall downward. We even know, understanding gravity and a few other things as well,

---

[2] L. Boltzmann, *Lectures on Gas Theory, 1896–1898*, trans. S. Brush (Berkeley, Calif : University of California Press, 1964), 446–7 (my italics).

why helium balloons, flames, etc. don't. A complete, coherent, and total explanation of all the phenomena we associated with the notion of 'down', associated in the sense of used to fix the very meaning, or at least reference, of 'down', is provided for us by the theory of gravitation.

We even know (although only vaguely to be sure) why it is that we can pick out the downward direction by an 'internal' sensation; why it is that we can know which way is down without ever observing an external falling object. The explanation has to do with the forces, once again gravitationally explained, exerted on the fluid of the inner ear. A demonstration of this and a full account is presumably a matter of some complexity, but we can rely on inference from the behaviour of simpler creatures. There are fish with sacs in their bodies with sand in the sac. Remove the sand and replace it with iron filings. Place a magnet over the fish tank and the fish swim upside-down. Surely it is something like that with us. In any case we don't doubt but that the ultimate physiological account of our inner apperception of down will refer ultimately to the effect of gravitational forces on some appropriate component of a bodily organ.

How should we describe appropriately the relationship between 'down' and 'the direction of the gradient of the gravitational field'? I am looking here, not for the ideal description in our ideally worked out semantico-metaphysics, but rather for the sorts of things we are, initially, intuitively inclined to say.

It wouldn't be strong enough to say that the downward direction is the direction of the gravitational gradient, for that would be true were it merely a happenstance that down and the direction of the gradient coincided. We feel, rather, that the downward direction is 'constituted' (whatever that means) by the direction of the gradient. Perhaps the correct locution is: Down (the downward relation itself) *is* (is identical to) the relation between points constituted by one's being deeper in the gravitational potential than the other. We *identify* the relation of a's being downward with respect to b with a's having a lower gravitational potential than b. (It is more complicated than that, of course, since b could be very remote from a, in which case we wouldn't talk that way if there were, for example, intervening regions of higher potential, but I am deliberately going to over-simplify grossly here.) Put this way the 'reduction' of the up–down to the gravitational relationship bears close analogy with substantival identifications as a means of theoretical reduction (water is H_2O, light waves are electromagnetic waves). But it is a property (relation) identification rather than one of substances.

I think that some would want to go further, arguing that the reduction established is sufficient to allow us to say that the very *meaning* of 'down' is given by the appropriate characterization of a relation in terms of the

gravitational gradient. Now, meaning is a notion as yet sufficiently uncon-
strained by a real theory as to allow us, with some plausibility, to say any
one of a number of different things. Emphasizing the connection of meaning
with criteria of applicability (verification procedures, operational definitions,
etc.) we would be inclined to say that although 'down' doesn't (or at least
didn't) *mean* 'the direction of the gravitational gradient', what was empirically discovered was that the downward relation was the relation gravitationally described. From this point of view there is a change of meaning which
has taken place when scientists, now fully aware of the gravitational account
of the up–down phenomena, begin to simply use 'down' to mean the local
direction of the gradient of the gravitational field.

Emphasizing, on the other hand, the association of meaning with reference,
in the manner of some recent semantic claims about proper names and natural
kind terms, we might, instead, be inclined to say things like: ' "Down" meant,
all along, the local direction of the gravitational gradient'. Of course it is still
a discovery on our part that gravity plays the explanatory role it does. From
this point of view we might even be tempted to say that prior to the full
understanding of the gravitational explanation of up–down phenomena.
people simply didn't understand what they meant by 'down'. And one will,
of course, now begin to claim that it is a necessary truth that the downward
direction is that of the gravitational gradient, allowing into one's scheme the
now familiar necessary a posteriori propositions which result from such a
'referentialist' semantics.[3] I do not wish to discuss any of the arguments for
or against such a view of meaning here, but only to emphasize, once again,
that in so far as a reduction of the up–down relationship to one characterized
in terms of the gravitational gradient is plausible at all, it is a reduction which
bears very striking analogies to the reductions by means of substantival
identification so familiar to us in other cases of intertheoretic reduction. It
bears an even closer resemblance to such property identifications as the
familiar (if abused) example of philosophers 'Temperature just is (is identical
to) mean kinetic energy of molecules'. Over-simplified as that claim may be,
the essence of what it is getting at is surely correct. 'Down just is (is identical
to) the direction of the gravitational gradient' seems a claim of the same order
and, if anything, probably in need of fewer qualifications and reservations
than are required in the thermodynamics-to-statistical mechanics case.

It is important to emphasize at this point the kind of reduction which the
one in question certainly is not. Perhaps no one would ever, in this context,
make the kind of mistake I am warning against here, but I believe that in the
context of the problem of the direction of time just such a confusion of kinds

---

[3] Cf. S Kripke, 'Naming and Necessity', in D. Davidson and G. Harman (eds.), *Semantics of
Natural Language* (Dordrecht: Reidel, 1972).

of reduction has played some role in muddying the waters. In saying that up–down reduced to the gravitational gradient relation we are not making a claim based upon a notion of priority of epistemic access. Such a claim is familiar to us in the claims that material object statements 'reduce' to sense-datum statements, spatio-temporal metric statements 'reduce' to statements about the local congruence of material-measuring instruments, etc. In the present case, unlike the ones just cited, there is no claim that our epistemic access to the up–down relationship is mediated through any sort of 'direct awareness' of the gravitational relationship; nor that some kind of hierarchy of epistemic immediacy tells us that up–down statements are, while initially thought of as inferred from gravitational statements, actually translatable into logical complexes of the gravitational-type statements. Instead the claim is just that the up–down relationship is found, by empirical research, to be identical with a more fundamental relationship characterizable in terms of the gravitational field. Down is the direction of the gravitational gradient as water is H_2O, light electromagnetic radiation and temperature mean kinetic energy. Not as tables are logical constructs out of sense data nor as non-local congruences are logical constructs out of spatio-temporally transported rods and clocks.

It will be useful at this point to say a little about one further aspect of the reduction of up–down to that of the gravitational gradient. Prior to under-standing the gravitational nature of down we intuitively viewed the down-ward direction in a global way: at every point the downward direction was parallel to the downward direction at every other point. (Of course this description of the situation is something of a travesty of the way in which the conceptual change occurred slowly over time. Aristarchus was well aware of the spherical nature of the earth and probably quite cognizant of the fact that down at Thebes was not parallel to down at Athens.) Recognizing the gravitational nature of the up–down relation we now realize clearly that what is down for us will most certainly not be parallel to what is down for someone at a different point on the earth's surface. We even understand that at some points of space there really won't be any downward direction at all.

Of course many ways are open to us to describe this. We can, if we wish, take 'down' to mean the direction of the gravitational gradient at the place we are located, identifying the downward direction elsewhere as the direction at that point parallel to our down. From that point of view Australians do, indeed, live their lives out upside-down. We might, to eliminate confusion, introduce a non-denumerable infinity of subscripted 'downs', 'down$_P$' refer-ring to the downward direction at the point referred to by the subscript. Then Australians live upside-down$_{USA}$ but, of course, right–side–up$_{AUST}$. Alterna-tively, and more elegantly, we can simply take 'down' as having an unequi-

vocal meaning but as functioning in the manner of a token-reflexive, at least
to the extent that:

(1) what is referred to as the downward direction by a speaker at one place
     is the direction of the gravitational gradient at that place;
(2) what is referred to as downward by a speaker at another place is the
     direction of the gravitational gradient at that place;
(3) and there is no reason whatever for thinking a priori that the referents
     of the two utterances of 'down' will be the same.

From this point of view, there is a clear sense in which the *sense* of 'down'
is the same for all speakers at all places.

5

I think it is clear that the entropic theory of temporal direction, if it is to be
plausible at all, should be viewed as a 'scientific' reduction motivated by an
empirical discovery of a property (relation) identification, and not as an
instance of the 'philosophical' reductions motivated in terms of a critical
analysis of the modes of epistemic access to the world available to us.
Perhaps this is obvious to many. But it hasn't always been obvious to me,
and at least some others have been misled. The following quote, for example,
is, perhaps, indicative of this confusion of modes of reduction. I think it is
appropriate here, even though it refers to a causal theory of the direction of
time, since, after all, the theory has been for many years an entropic theory
of temporal direction rather than a causal theory.

It is sometimes suggested that the direction of time and causation are linked because
the direction of time is itself to be analysed in terms of causation. But, at least as
conceptual analysis, this must be wrong. We can think of events' succeeding one
another in time even if there are no causal links between any of them, let alone between
the members of each pair of which one is earlier than the other. Moreover, *our concept
of the direction of time is based on a pretty simple, immediate, experience of one event's
following straight after another,* or of a process going on—say of something's moving—with a later phase following an earlier one. It might be, of course, that our having
such experiences is somehow dependent upon causally asymmetrical processes going
on inside us—we might have internally causally controlled unconscious temporal direction indicators—but even if this were so it would not mean that our concept of time
direction was analysable into that of causal direction. *Our experience of earlier and
later, on which our concept of time direction is based, itself remains primitive,* even if
it has some unknown causal source.[4]

[4] J. Mackie, 'Causal Asymmetry in Concept and Reality', paper presented at the 1977 Oberlin
Colloquium in Philosophy, 1 (my italics). For other expressions of scepticism about the entropic
theory, see J. Earman, 'An Attempt to Add a Little Direction to "The Problem of the Direction of

But if the entropic theorist has in mind reduction of the 'scientific' kind, then nothing in the way of immediate, simple experience of earlier and later events, or ongoing processes, nor any reference to an ability to imagine (think of) events being temporally ordered without being entropically related will refute the claim, meant in this sense, that the later-than relation is (is identical to) some relation characterized in terms of entropy, nor even that, in the senses of meaning we noted above, in some sense 'later than' *means* 'bears some appropriate entropically characterizable relation to'.

Since the two notions of concept reduction I have been discriminating are easily confused in general, it isn't too much of a surprise that we have not always been clear which sense of reduction is intended by the entropic theory of temporal direction. But I think that some of the very arguments used by entropic theorists have tended to ingrain the confusion. For example, entropic theorists frequently ask us to consider how we would distinguish a film of events run in the proper order from the film run in reverse order, pointing out to us that the discrimination can only be done (or, rather, so it is claimed) when entropic features of the world are present, and that it is by means of the expected dissipation of order into disorder that we make the judgement about whether the film is being run in the correct direction. If this is meant only to show us that the entropic features of the world are, at least, the most prominent which are asymmetric in time order and, hence, the prime candidates for a reduction of the scientific kind, then it is harmless. But it is easy to slip from this argument into the dubious claim that we judge the time order of events in the actual world by inference from apprehension of ordering of states in respect to entropy. As Mackie and others have pointed out, this is indeed dubious. But the dubiousness of that latter claim is an argument only against the 'philosophical' theory of the reduction of time order to entropy. In no way would it vitiate a reductive claim of the 'scientific' kind.

Again consider Reichenbach's transition from a causal to an entropic direction. If what is being said there is that the only relevant causal notion is causal connectibility, that this is temporally symmetric, and hence not a suitable candidate for a reduction basis for the relation of temporal order, then it is a point relevant to reductions of the identificatory kind. But it is easy to read the argument as saying that the causal theory won't do because we must be able to empirically *determine* which of a causally connected pair of events is cause and which effect in order to make the reduction go through, and that this determination requires first *knowing* the time order of events, and that this makes causation unsuitable as a reduction basis for temporal

Time" ', *Philosophy of Science*, 41 (1974): 15–47, and my book *Space, Time, and Spacetime* (Berkeley, Calif.: University of California Press, 1974), 404–11. Most important, see A. Eddington, *The Nature of the Physical World* (Cambridge: Cambridge University Press, 1928), ch. 5, 'Becoming'.

direction because it lacks the necessary epistemic independence and primacy. But this latter argument, once again, suggests that it is the epistemically motivated kind of reduction which the theorist has in mind. If he then offers an entropic theory as the substitute for the causal, one is misled into thinking that the theory too is an attempted reduction of the 'philosophical' sort.

There is also the fact that Reichenbach presents the entropic theory as part of a general reductivist account of space-time. Entropy is to fix one last part of space-time structure, the past–future distinction, after the rest of space-time, in particular its topology, including its temporal topology, has already been 'reduced' to non-prima facie spatio-temporal notions. In particular, the space-time topology is supposed to be reduced to the *causal* structure of the world.

Now I think that a kind of 'scientific' identificatory reduction of space-time topology to causal order could be argued for. For example, the recent suggestions of reducing space-time structure to some kind of algebraic relationship among quantum events might be viewed as a reductionist move of this kind. But I think that the causal theory of space-time topology which Reichenbach offers is, rather, motivated by, and formulated in terms more appropriate to, an epistemically generated type of 'philosophical' reduction. If this is correct we can see why one would easily be misled into thinking that the entropic account of time direction was also supposed to be a reduction of this latter kind.[5]

6

But if the entropic theory of the direction of time is supposed to be a scientific reduction we must ask whether or not it is successful. Is the connection of entropy with time order like that of asymmetric weak interaction processes with left and right, merely a correlation (lawlike or *de facto*), or is the case rather like that of gravitation and up and down, where we feel it is at least appropriate to say that the up–down relation is identical to the gravitationally characterized relation, and where we are even tempted (at least on some theories of meaning) to say that 'down' means 'in the direction of the gradient of the gravitational field'?

That question I hardly intend to try and answer here. What is needed is a full-fledged attempt to try and account for all the processes we normally (pre-scientifically) take to mark out the direction of time, including our internal 'direct' sense of temporal order, in terms of a single, unified account

---

[5] On the causal theory of space-time topology as an instance of philosophical reduction, see my article 'What Might Be Right about the Causal Theory of Time', *Synthese*, 35 (1977): 155–71.

which invokes the relation of difference in entropy and accounts for all these phenomena in terms of an identification of the time order relationship with some relationship among events characterizable (at least in part as we shall see) in entropic terms, and which does not invoke time order itself as a primitive in the characterization. Despite Reichenbach's heroic efforts in this direction, I think we can all agree that such an account is not yet available to us.[6] But, of course, Reichenbach's efforts, from this point of view of the nature of the entropic theory, are at least efforts in the right direction. We must explain, entropically, why causes precede their effects (at least usually); why we have records of the past and not the future; why we know and believe so much more about the past than the future, and believe and know about them in such very different ways; why we feel we can change the future but not the past; why we have such a different emotional attitude toward the future than we do to the past ('Thank God that's over!'); why we take the past to have determinate reality and the future to exist, if at all, merely as 'pure potentiality'; and, finally, why we have direct, immediate, non-inferential knowledge of the time order of events (internal and external) with which we are directly acquainted (in Russell's sense).

While many of Reichenbach's arguments in these directions are brilliantly imaginative and suggestive, I do not believe that I will be taken to be disrespectful if I assert here that they are, to many of us, far from conclusive. They serve as brilliant suggestions toward a theory, but the theory we will ultimately be given by the entropist as highly confirmative of his reductive claim is still in the future. Here I wish only to make a few rather general remarks about the entropic programme, some of the difficulties it faces, and why at least some suggested objections to it are not really devastating to its aim.

1. At least part of the problem in trying to establish the entropic theory is the rather vague grasp we have on many of the notions to be accounted for entropically in the reduction. Compare asking: 'What is a causal relation?' 'What is a record or trace of an event?', etc. with asking: 'What is a falling object?' In the gravitational theory of down at least we have, prior to the reduction, a pretty good idea of what it is that the gravitational theory must account for. In the entropic theory of time direction we don't have a very clear idea at all. Of course, it may very well be claimed by the entropic theorist that it is only in the context of the reduction that our ideas of what it is that must be accounted for will become clear. I think that Reichenbach has this in mind. For example, only when we understand the role played by

    [6] See H. Reichenbach, *The Direction of Time* (Berkeley, Calif.: University of California Press, 1956).

entropic features of the world in our pre-scientific conceptual scheme will we really begin to understand our pre-analytically felt, but very poorly understood, intuition that causal efficacy proceeds from past to present and thence to future.

2. I have deliberately avoided any attempt at saying exactly what relationship among events, characterized entropically, is the one with which the 'later than' relationship is to be identified. It is clear that this identification will be one of some subtlety. In the case of the up–down relationship the identification is fairly simple-minded. If b is downward from a then there is a gravitational potential difference between them determined by the value of the potential at the two points. That plus some facts about the gravitational potential at intermediate points is enough to fix the appropriate 'gravitational' relationship in our reduction basis.

But the temporal asymmetry case is trickier. First of all there is the fact that a later state of even an isolated system can very well be one of lower entropy than an earlier state. We must take account of the fact that the association of entropy order with time order is supposed to be only statistical. Secondly, there is the fact that we take the time-ordering relationship to be more pervasive than that of entropic order, in the sense that there can be a later-than relationship holding between events where no obvious characterization of the events as states of affairs of a system with different entropy is at all possible.

Now one solution to this would be to postulate the existence of a 'time potential' with a gradient in the timelike direction, making the time order case look far more like the up–down case than even I have maintained it to be. This is Weingard's suggestion in a 1977 article.[7] I think this is the wrong way to go. I don't deny that in some possible world that is how things could turn out to be, with the existence of such a time-ordering vector field as a 'real physical field' whose existence is ultimately explanatory of the familiar asymmetries of the world in time. It is just that we have no reason whatever to believe that in this world there is any such field. The usual statistical mechanical explanations of the asymmetric behaviour of systems in time invoke no such fundamental field. Granted we frequently do find the statistical mechanical explanations unsatisfying, and many have the feeling that at the present state of our understanding some matters of fundamental importance have yet to be uncovered. But few physicists would presently accept as a plausible explanation the existence of such a fundamental time-ordering field as the underlying 'missing link' in attempting to offer a full explanation of the asymmetry of the world in time. If the theory of time direction is

[7] R. Weingard, 'Space-Time and the Direction of Time', *Noûs*, 11 (1977): 119–32.

supposed to be a scientifically established identificatory reduction of the later-than relation to some other more fundamental relation, then it must be established by the real science of the world as it actually is. A possible reduction which would be satisfactory in some possible, but non-actual, world is of no help to us.

Nor need we invoke such a pseudo-field in order to have an adequate account. One direction in which to move is again available to us in Reichenbach. The whole entropic theory presupposes an underlying theory of time order—the full topology of time (or space-time)—with the only intuitive feature removed being that of the past–future asymmetry. Of nearby pairs of pairs of events (nearby to avoid the possibility of non-temporally orientable space-times) we can ask if d is in the same time direction from c as b is from a. If we can then establish the 'laterness' of, say, b to a on the basis of entropic considerations, we can 'project' this time order on to the $c–d$ pair, taking d as later than c, even if none of the relevant entropic considerations appear in the $c–d$ case. Actually, of course, the detailed theory might be much more complicated than that, making reference, possibly, to multitudes of systems and, where entropies can be assigned to temporally distinct states of isolated systems, to entropic difference 'parallel' for the overwhelming majority of them. Then the past–future time direction is taken as being fixed by this majority of systems, lower entropy states being earlier than higher entropy states, and thence 'projected' by local comparability of time order to all pairs of temporally related events. For our present purposes the details are inessential.

3. What I have suggested above suggests an approach to the entropic account which offers a 'definition' for time direction in terms of the entropic behaviour of branch systems, in Reichenbach's terminology.[8] I might say something here about the relevance of the branch system notion to the overall account. With what entropic feature of the world do we wish to identify the time direction? Not, if Boltzmann's overall approach is correct, with the entropic relationship between states of the universe as a whole (assuming that such a notion as entropy for the universe as a whole is well defined, a powerful and dubious assumption). More plausible would be an identification of the later-than relation at a place-time with the appropriate entropic relation among the states of the 'single world' during the 'aeon' containing the place-time. This, indeed, might be the right direction for the entropic theorist to go, rather than that outlined above.

Why need we invoke the branch systems of Reichenbach? If we were holding to an epistemically motivated 'philosophical' reduction, the answer

---

[8] See Reichenbach, *The Direction of Time*, 118–43.

would be obvious. We have, certainly, no epistemic access of a direct sort to the total entropy even of our 'single world'. But, perhaps, we do to local temporarily isolated systems. So we observe them, and the entropic relationships among their states, and 'infer' time order from these relationships. The reduction then consists in replacing this 'inference' with a 'co-ordinative definition' in the familiar way.

But I have been maintaining that it is not this kind of reduction which the entropic theorist is really after. What then is the role of branch systems and their entropically characterizable states? I think it is that, in our explanations of the various phenomena characteristic of the asymmetry of the world in time intuitively associated with the time order of the world in our pre-scientific picture, the branch systems and their states will have to be invoked. Even if ultimately in the explanation of these asymmetries we refer to the entropic behaviour of our 'single world' during its present 'aeon', the explanation will invoke at an intermediate stage some account of how this entropic asymmetry gives rise to the entropic asymmetries of the branch systems, and will then use these 'small' entropic asymmetries to account for the familiar asymmetries of causation, knowledge, traces, etc. and to account for our immediate internal sense of the time order of our own experiences.

Whether the entropic theorist will then want to identify the later-than relationship with a relationship entropically characterized among the total states of the 'single world' or, instead, with some complex 'majority rule' relationship among states of sets of branch systems I do not know. I think we would need more detail about the nature of the entropic theory to decide this. Or, perhaps, he has a choice and there is an element of arbitrariness in the identification he asserts.

4. We saw that, in the gravitational reduction of the concept of up and down to that of the direction of the gravitational gradient, it was no argument against the account at all that at different places in space the downward direction could vary. The same holds true with the entropic theory of time direction. Whether or not Boltzmann is right that there are at a given time 'single worlds' with their time orders oppositely directed, or a single 'single world' which at different 'aeons' has its time order in the reverse direction, this is certainly a possible state of affairs on the entropic account. And nothing about this state of affairs makes the entropic theory in any way less plausible.

Once again we have a choice of at least two ways of describing the situation. We can take the future direction of time as fixed by the entropic relations among states of our 'single world' in our 'aeon' and speak of entropy as going the 'wrong way' in time in the counter-directed worlds. Or, less parochially, we could take the past–future relationship to be quasi-indexical, letting 'future' refer to that direction in time at a space-time point which is the direction bearing the appropriate entropic feature in that 'single world' during that 'aeon'. None of this is incompatible with the earlier remarks that the entropic theorist might wish to use *local* comparability of time order to project the past–future relationship from some system in his 'single world' to others.[9]

7

At this point my already very sketchy and somewhat vague paper is going to become even less the presentation of a polished, finished account. For I am here going to suggest that, at a new level, some of the standard objections to the entropic account may reappear, even if that account is interpreted in its most plausible form as an account of a 'scientific' identificatory reduction.

At some point the reductive programmes of the naturalistic sort which proceed by identificatory reductions of substances and properties to those more 'scientifically' fundamental, and the reductive programmes of the philosophical sort which proceed by 'conceptual analysis' of propositions in terms of a critical examination of the total class of propositions which could serve as epistemic warrant for them, must be reconciled. One could, of course, reject the latter kind of reductionism altogether as spurious but I don't think that we can do this without at the same time rejecting some of the deepest and best-accepted portions of our recent scientific progress; for, I would allege, much of the transition from space and time to relativistic space-time proceeds by just such an epistemically motivated 'reductionist' critique. I won't argue this here, but only try to show how one aspect of 'scientific' reductions introduces, in the particular case of the entropic account of the direction of time, some special difficulties closely related to the problem of working together these two kinds of reductive analysis.

A familiar concomitant of identificatory reductions is the 'secondarizing' of properties. Tables are arrays of atoms. But what about the 'immediately sensed' properties of macroscopic tables? Are they properties of arrays of atoms? Arrays of atoms are, in some sense, discontinuous; but what of the sensed, continuous colour patch that a table presents to my awareness? One solution (maybe not the only and maybe not the best) is to remove properties from the table (except for leaving a residuum of them as powers or disposi-

---

[9] The arguments here are in reply to an argument of Earman's, especially to his invocation of what he calls the Principle of Precedence. See Earman, 'An Attempt to Add a Little Direction to "The Problem of the Direction of Time"', 21–3.

tions) and reclassify them as secondary qualities of sense data or, perhaps, of the sensing perceiver (who is appeared to reddishly, etc.)

Temperature is mean kinetic energy of molecules. But what of the felt quality we first used to discriminate hotter from colder objects? Easy, make it a secondary quality 'in the mind' of the perceiver. Whenever we propose an identificatory reduction of some entity or property, initially identified by us by a 'direct apprehension', to some other entities or properties in the world, there is at least the temptation to strip off from the object the original identifying feature and place it 'in the mind' as a secondary quality related to the reducing property in the world only as the causal effect of that property's acting, by means of the sensory apparatus, on the 'mind'. I'm not saying that this is the only direction in which to go nor that it is the right one; only that it is a persistent, common move and one intuitively hard to resist.[10]

Now we take the later-than relation to be a relation in the world characterizable in entropic terms. But what of the 'pretty simple, immediate, experience of one event's following straight after another'? Our temptation is, I think, once again to dissociate the immediately sensed, directly apprehended, 'later-than-ness' of events from the time order of events in the world, making it into a feature only of events 'in the mind'.

But now we see why many who would easily accept the claim that tables are, in fact, arrays of atoms, and that temperature is, in fact, mean kinetic energy of molecules, will balk at the claim that later-than-ness is an entropically characterizable relation among events in the world. We feel that time order is something that holds of events in the world and the events of inner experience as well. Since Kant we have been familiar with the claim that space is the manifold of experience of outer objects and time of both inner and outer awareness. But it is the same time which relates outer events and which relates events 'in the mind'. And if outer events are later than one another, are they not later than one another in exactly the same sense that inner experiences occur in the asymmetric order of time? And if I directly experience this order among events in my inner mental life, mustn't I identify that relation with the real later-than relation among events in the world? If these events in the world are also related by some relation entropically characterizable, mustn't that be viewed as an empirically established correlation with time order then? And isn't it then true that there is no more plausibility in identifying the later-than relation with the entropic relation than there is in identifying left-handedness with some feature of an object characterized in terms of the behaviour of weak interactions?

---

[10] See Lawrence Sklar, 'Types of Inter-theoretic Reduction', *British Journal for the Philosophy of Science*, 18 (1967): 122–3.

Notice the difference here from the gravitational case. Our inner experiences are not, really, up and down from one another. No harm then in dissociating our inner experience of down from the real down-ness relation in the world and then identifying the latter with a relation characterized in terms of the gradient of the gravitational potential. But inner events are *really* later and earlier than one another and our 'pretty simple, immediate experience' of this relation cannot with impunity be detached as merely a causally induced secondary quality not properly thought of as a direct experience of the real afterness relation which exists in the world only as an entropically characterizable relation.

The following quote from Eddington suggests that it is something like this argument which is at the root of many of the strongly felt but not very well expressed objections to the plausibility of the entropic theory. It is important to note that this quote is from one of the earliest expositors of the entropic theory of time direction as I have described it.

In any attempt to bridge the domains of experience belonging to the spiritual and physical sides of our nature, Time occupies the key position. I have already referred to its dual entry into our consciousness—through the sense organs which relate it to the other entities of the physical world, and directly through a kind of private door into the mind. . . . Whilst the physicist would generally say that the matter of this familiar table is *really* a curvature of space, and its colour is really electromagnetic wavelength, I do not think he would say that the familiar moving on of time is *really* an entropy-gradient. . . . Our trouble is that we have to associate two things, both of which we more or less understand, and, so as we understand them, they are utterly different. It is absurd to pretend that we are in ignorance of the nature of organization in the external world in the same way that we are ignorant of the intrinsic nature of potential. It is absurd to pretend that we have no justifiable conception of 'becoming' in the external world. That dynamic quality—that significance which makes a development from future to past farcical—has to do much more than pull a trigger of a nerve. It is so welded into our consciousness that a moving on of time is a condition of consciousness. We have direct insight into 'becoming' which sweeps aside all symbolic knowledge as on an inferior plane. If I grasp the notion of existence because I myself exist, I grasp the notion of becoming because I myself become. It is the innermost Ego of all which *is* and *becomes*.[11]

I don't pretend to understand all that Eddington is saying here, nor to be able to give a really coherent version of my own arguments above. I do think, however, that it is very clear that our ultimate view of the world will require a subtle and careful weaving together of the naturalistic reduction of science which proceeds by theoretical identification with the conceptual reduction of philosophy which proceeds by epistemic analysis. Until we have such a systematic overall account I think that the ultimate status of an entropic theory of time order will be in doubt.

---

[11] Eddington, *The Nature of the Physical World*, 91–7.

VII

BRINGING ABOUT THE PAST

MICHAEL DUMMETT

I observe first that there is a genuine sense in which the causal relation has a temporal direction: it is associated with the direction earlier-to-later rather than with the reverse. I shall not pause here to achieve a precise formulation of the sense in which this association holds; I think such a formulation can be given without too much difficulty, but it is not to my present purpose to do this. What I do want to assert is the following: so far as I can see, this association of causality with a particular temporal direction is not merely a matter of the way we speak of causes, but has a genuine basis in the way things happen. There is indeed an asymmetry in respect of past and future in the way in which we describe events when we are considering them as standing in causal relations to one another; but I am maintaining that this reflects an objective asymmetry in nature. I think that this asymmetry would reveal itself to us even if we were not *agents* but mere *observers*. It is indeed true, I believe, that our concept of cause is bound up with our concept of intentional action: if an event is properly said to cause the occurrence of a subsequent or simultaneous event, I think it necessarily follows that, if we can find any way of bringing about the former event (in particular, if it is itself a voluntary human action), then it must make sense to speak of bringing it about *in order that* the subsequent event should occur. Moreover, I believe that this connection between something's being a cause and the possibility of using it in order to bring about its effect plays an essential rôle in the fundamental account of how we ever come to accept causal laws: that is, that we could arrive at any causal beliefs only by beginning with those in which the cause is a voluntary action of ours. Nevertheless, I am inclined to think that we could have some kind of concept of cause, although one differing from that we now have, even if we were mere observers and not agents at all—a kind of intelligent tree. And I also think that even in this case the asymmetry of cause with respect to temporal direction would reveal itself to us.

Michael Dummett, 'Bringing About the Past'. First published in the *Philosophical Review*, 73 (1964): 338–59. The text as reprinted here includes some small changes made by the author when this essay was reprinted in his *Truth and Other Enigmas* (London: Duckworth, 1978).

To see this, imagine ourselves observing events in a world just like the actual one, except that the order of events is reversed. There are indeed enormous difficulties in describing such a world if we attempt to include human beings in it, or any other kind of creature to whom can be ascribed intention and purpose (there would also be a problem about memory). But, so far as I can see, there is no difficulty whatever if we include in this world only plants and inanimate objects. If we imagine ourselves as intelligent trees observing such a world and communicating with one another, but unable to intervene in the course of events, it is clear that we should have great difficulty in arriving at causal explanations that accounted for events in terms of the processes which had *led up to* them. The sapling grows gradually smaller, finally reducing itself to an apple pip; then an apple is gradually constituted around the pip from ingredients found in the soil; at a certain moment the apple rolls along the ground, gradually gaining momentum, bounces a few times, and then suddenly takes off vertically and attaches itself with a snap to the bough of an apple tree. Viewed from the standpoint of gross observation, this process contains many totally unpredictable elements: we cannot, for example, explain, by reference to the conditions obtaining at the moment when the apple started rolling, why it started rolling at that moment or in that direction. Rather, we should have to substitute a system of explanations of events in terms of the processes that led back to them from some subsequent moment. If through some extraordinary chance we, in this world, could consider events from the standpoint of the microscopic, the unpredictability would disappear theoretically ('in principle') although not in practice; but we should be left—so long as we continued to try to give causal explanations on the basis of what leads up to an event—with inexplicable coincidences. 'In principle' we could, by observing the movements of the molecules of the soil, predict that at a certain moment they were going to move in such a way as to combine to give a slight impetus to the apple, and that this impetus would be progressively reinforced by other molecules along a certain path, so as to cause the apple to accelerate in such a way that it would end up attached to the apple tree. But not only could we not make such predictions in practice: the fact that the 'random' movements of the molecules should happen to work out in such a way that all along the path the molecules always happened to be moving in the same direction at just the moment that the apple reached that point, and, above all, that these movements always worked in such a way as to leave the apple attached to an *apple* tree and not to any other tree or any other object—these facts would cry out for explanation, and we should be unable to provide it.

I should say, then, that, so far as the concept of cause possessed by mere observers rather than agents is concerned, the following two theses hold: (i) the world is such as to make appropriate a notion of causality associated with the earlier-to-later temporal direction rather than its reverse; (ii) we can conceive of a world in which a notion of causality associated with the opposite direction would have been more appropriate and, so long as we consider ourselves as mere observers of such a world, there is no particular conceptual difficulty about the conception of such a backwards causation. There are, of course, regions of which we are mere observers, in which we cannot intervene: the heavens, for example. Since Newton, we have learned to apply the same causal laws to events in this realm; but in earlier times it was usually assumed that a quite different system of laws must operate there. It *could* have turned out that this was right; and then it could also have turned out that the system of laws we needed to explain events involving the celestial bodies required a notion of causality associated with the temporal direction from later to earlier.

When, however, we consider ourselves as agents, and consider causal laws governing events in which we can intervene, the notion of backwards causality seems to generate absurdities. If an event C is considered as the cause of a preceding event D, then it would be open to us to bring about C in order that the event D should have occurred. But the conception of doing something in order that something else should have happened appears to be intrinsically absurd: it apparently follows that backwards causation must also be absurd in any realm in which we can operate as agents.

We can affect the future by our actions: so why can we not by our actions affect the past? The answer that springs to mind is this: you cannot *change* the past; if a thing has happened, it has happened, and you cannot make it not to have happened. This is, I am told,[1] the attitude of orthodox Jewish theologians to retrospective prayer. It is blasphemous to pray that something should *have* happened, for, although there are no limits to God's power, He cannot do what is logically impossible; it is logically impossible to alter the past, so to utter a retrospective prayer is to mock God by asking Him to perform a logical impossibility. Now I think it is helpful to think about this example, because it is the only instance of behaviour, on the part of ordinary people whose mental processes we can understand, designed to affect the past and coming quite naturally to us. If one does not think of this case, the idea of doing something in order that something else should previously have happened may seem sheer raving insanity. But suppose I hear on the radio that a ship has gone down in the Atlantic two hours previously, and that there were a few survivors: my son was on that ship, and I at once utter a prayer that he should have been among the survivors, that he should not have

---

[1] By Professor G. Kreisel.

drowned; this is the most natural thing in the world. Still, there are things which it is very natural to say which make no sense; there are actions which can naturally be performed with intentions which *could* not be fulfilled. Are the Jewish theologians right in stigmatizing my prayer as blasphemous?

They characterize my prayer as a request that, if my son has drowned, God should make him not have drowned. But why should they view it as asking anything more self-contradictory than a prayer for the future? If, before the ship set sail, I had prayed that my son should make a safe crossing, I should not have been praying that, if my son was going to drown, God should have made him not be going to drown. Here we stumble on a well-known awkwardness of language. There is a use of the future tense to express present tendencies: English newspapers sometimes print announcements of the form 'The marriage that was arranged between X and Y will not now take place'. If someone did not understand the use of the future tense to express present tendencies, he might be amazed by this 'now'; he might say, 'Of course it *will* not take place *now:* either it *is* taking place *now*, or it *will* take place *later*'. The presence of the 'now' indicates a use of the future tense according to which, if anyone had said earlier, 'They are going to get married,' he would have been right, even though their marriage never subsequently occurred. If, on the other hand, someone had offered a bet which he expressed by saying, 'I bet they will not be married on that date,' this 'will' would normally be understood as expressing the *genuine* future tense, the future tense so used that what happens on the future date is the decisive test for truth or falsity, irrespective of how things looked at the time of making the bet, or at any intervening time. The future tense that I was using, and that will be used throughout this paper, is intended to be understood as this genuine future tense.

With this explanation, I will repeat: when, before the ship sails, I pray that my son will make the crossing safely, I am not praying that God should perform the logically impossible feat of making what will happen not happen (that is, not be-going-to happen); I am simply praying that it will not happen. To put it another way: I am not asking God that He should now make what is going to happen not be going to happen; I am asking that He *will* at a future time make something not to happen at that time. And similarly with my retrospective prayer. Assuming that I am not asking for a miracle—asking that, if my son has died, he should now be brought to life again—I do not have to be asking for a logical impossibility. I am not asking God that, even if my son has drowned, He should *now* make him not to have drowned; I am asking that, at the time of the disaster, He should then have made my son not to drown at that time. The former interpretation would indeed be required if the list of survivors had been read out over the radio, my son's name had not been on it, and I had not envisaged the possibility of a mistake on the part of the news service: but in my ignorance of whether he was drowned or not, my prayer will bear another interpretation.

But this still involves my trying to affect the past. On this second interpretation, I am trying by my prayer *now* to bring it about that God made something not to happen: and is not this absurd? In this particular case, I can provide a rationale for my action—that is why I picked this example—but the question can be raised whether it is not a bad example, on the ground that it is the only kind for which a rationale *could* be given. The rationale is this. When I pray for the future, my prayer makes sense because I know that, at the time about which I am praying, God will remember my prayer, and may then grant it. But God knows everything, both what has happened and what is going to happen. So my retrospective prayer makes sense, too, because, at the time about which I am praying, God knew that I was going to make this prayer, and may then have granted it. So it seems relevant to ask whether foreknowledge of this kind can meaningfully be attributed only to God, in which case the example will be of a quite special kind, from which it would be illegitimate to generalize, or whether it could be attributed to human beings, in which case our example will not be of purely theological interest.

I have heard three opinions expressed on this point. The first, held by Russell and Ayer, is that foreknowledge is simply the mirror image of memory, to be explained in just the same words as memory save that 'future' replaces 'past', and so forth, and as such is conceptually unproblematic: we do not have the faculty but we perfectly well might. The second is a view held by a school of Dominican theologians. It is that God's knowledge of the future should be compared rather to a man's knowledge of what is going to happen, when this lies in his intention to make it happen. For example, God knows that I am going to pray that my son may not have drowned because He is going to make me pray so. This leads to the theologically and philosophically disagreeable conclusion that everything that happens is directly effected by God, and that human freedom is therefore confined to wholly interior movements of the will. This is the view adopted by Wittgenstein in the *Tractatus*, and there expressed by the statement 'The world is independent of my will.' On this view, God's foreknowledge is knowledge of a type that human beings do have; it would, however, be difficult to construct a non-theological example of an action intelligibly designed to affect the past by exploiting this alleged parallelism. The third view is one of which it is difficult to make a clear sense. It is that foreknowledge is something that can be meaningfully ascribed only to God (or perhaps also to those He directly inspires, the prophets; but again, perhaps these would be regarded not as themselves possessing this knowledge, but only as the instruments of its expression). The ground for saying this is that the future is not something of which we could, but merely do not happen to, have knowledge; it is not, as it were, *there* to be known. Statements about the future are, indeed, either-true-or-false; but they do not yet have a particular one of these two truth-values. They have present truth-or-falsity, but they do not have present truth or present falsity, and so they *cannot* be known: there is not really anything to be known. The non-theological part of this view seems to me to rest on a philosophical confusion; the theological part I cannot interpret, since it appears to involve ascribing to God the performance of a logical impossibility.

We saw that retrospective prayer does not involve asking God to perform the logically impossible feat of changing the past, any more than prayer for the future involves asking Him to change the future in the sense in which that is logically impossible. We saw also that we could provide a rationale for retrospective prayer, a rationale which depended on a belief in God's foreknowledge. This led us to ask if foreknowledge was something which a man could have. If so, then a similar rationale could be provided for actions designed to affect the past, when they consisted in my doing something in order that someone should have known that I was going to do it, and should have been influenced by this knowledge. This enquiry, however, I shall not pursue any further. I turn instead to more general considerations: to consider other arguments designed to show an intrinsic absurdity in the procedure of attempting to affect the past—of doing something in order that something else should have happened. In the present connection I remark only that, if there is an intrinsic absurdity in *every* procedure of this kind, then it follows indirectly that there is also an absurdity in the conception of foreknowledge, human or divine.

Suppose someone were to say to me, 'Either your son has drowned or he has not. If he has drowned, then certainly your prayer will not (cannot) be answered. If he has not drowned, your prayer is superfluous. So in either case your prayer is pointless: it cannot make any *difference* to whether he has drowned or not.' This argument may well appear quite persuasive, until we observe that it is the exact analogue of the standard argument for fatalism. I here characterize fatalism as the view that there is an intrinsic absurdity in doing something in order that something else should subsequently happen; that any such action—that is, any action done with a further purpose—is necessarily pointless. The standard form of the fatalist argument was very popular in London during the bombing. The siren sounds, and I set off for the air-raid shelter in order to avoid being killed by a bomb. The fatalist argues, 'Either you are going to be killed by a bomb or you are not going to be. If you are, then any precautions you take will be ineffective. If you are not, all precautions you take are superfluous. Therefore it is pointless to take precautions.' This belief was extended even to particular bombs. If a bomb was going to kill me, then it 'had my number on it', and there was no point in my attempting to take precautions against being killed by *that* bomb; if it did not have my number on it, then of course precautions were pointless too. I shall take it for granted that no one wants to accept this argument as cogent. But the argument is formally quite parallel to the argument supposed to show that it is pointless to attempt to affect the past; only the tenses are different. Someone may say, 'But it is just the difference in tense that makes the difference between the two arguments. Your son has either *already* been drowned or else *already* been saved; whereas you haven't *yet* been killed in the raid, and you haven't *yet* come through it.' But this is just to reiterate that the one argument is about the past and the other about the future: we want to know what, if anything, there is *in* this fact which makes the one valid, the other invalid. The best way of asking this question is to ask, 'What refutation is there of the fatalist argument, to which a quite parallel refutation of the argument to show that we cannot affect the past could not be constructed?'

Let us consider the fatalist argument in detail. It opens with a tautology, 'Either you are going to be killed in this raid or you are not'. As is well known, some philosophers have attempted to escape the fatalist conclusion by faulting the argument at this first step, by denying that two-valued logic applies to statements about future contingents. Although this matter is worth investigating in detail, I have no time to go into it here, so I will put the main point very briefly. Those who deny that statements about future contingents need be either true or false are under the necessity to explain the meaning of those statements in some way; they usually attempt to do so by saying something like this: such a statement is not true or false now, but *becomes* true or false at the time to which it refers. But if this is said, then the fatalist argument can be reconstructed by replacing the opening tautology by the assertion 'Either the statement "You will be killed in this raid" is going to become true, or it is going to become false'. The only way in which it can be consistently maintained not only that the law of excluded middle does not hold for statements about the future, but that there is no other logically necessary statement which will serve the same purpose of getting the fatalist argument off the ground, is to deny that there is, or could be, what I called a 'genuine' future tense at all: to maintain that the only intelligible use of the future tense is to express present tendencies. I think that most people would be prepared to reject this as unacceptable, and here, for lack of space, I shall simply assume that it is. (In fact, it is not quite easy to refute someone who consistently adopts this position; of course, it is always much easier to make out that something is not meaningful than to make out that it is.) Thus, without more ado, I shall set aside the suggestion that the flaw in the fatalist argument lies in the very first step.

The next two steps stand or fall together. They are: 'If you are going to be killed in this raid, you will be killed whatever precautions you take' and 'If you are not going to be killed in this raid, you will not be killed whatever precautions you neglect'. These are both of the form 'If p, then if q then p'; for example, 'If you *are* going to be killed, then you will be killed even if you take precautions'. They are clearly correct on many interpretations of 'if'; and I do not propose to waste time by enquiring whether they are correct on 'the' interpretation of 'if' proper to well-instructed users of the English language. The next two lines are as follows: 'Hence, if you are going to be killed in the raid, any precautions you take will be ineffective' and 'Hence, if you are not going to be killed in the raid, any precautions you take will have been superfluous'. The first of these is indisputable. The second gives an appearance of sophistry. The fatalist argues from 'If you are not going to be killed, then you won't be killed even if you have taken no precautions' to 'If you are not going to be killed, then any precautions you take will have been superfluous'; that is, granted the truth of the statement 'You will not be killed even if you take no precautions', you will have no motive to take precautions; or, to put it another way, if you would not be killed even if you took no precautions, then any precautions you take cannot be considered as being effective in bringing about your survival—that is, as effecting it. This employs a well-known principle. St Thomas, for instance, says it is a condition of ignorance to be an excuse for having done wrong that, if the person had not suffered from the ignorance, he would not have committed the wrongful act in question. But we want to object that it may be just the precautions that I am going to take which save me from being killed; so it cannot follow from the mere fact that I am not going to be killed that I should not have been going to be killed even if I had not been going to take precautions. Here it really does seem to be a matter of the way in which 'if' is understood; but, as I have said, I do not wish to call into question the legitimacy of a use of 'if' according to which '(Even) if you do not take precautions, you will not be killed' follows from 'You will not be killed'. It is, however, clear that, on any use of 'if' on which this inference is valid, it is possible that both of the statements 'If you do not take precautions, you will be killed' and 'If you do not take precautions, you will not be killed' should be true. It indeed follows from the truth of these two statements together that their common antecedent is false; that is, that I am in fact going to take precautions. (It may be held that on a, or even the, use of 'if' in English, these two statements cannot both be true; or again, it may be held that they can both be true only when a stronger consequence follows, namely, that not only am I as a matter of fact going to take precautions, but that I could not fail to take them, that it was not in my power to refrain from taking them. But, as I have said, it is not my purpose here to enquire whether there are such uses of 'if' or whether, if so, they are important or typical uses.) Now let us say that it is correct to say of certain precautions that they are capable of being effective in preventing my death in the raid if the two conditional statements are true that, if I take them, I shall not be killed in the raid, and that, if I do not take them, I shall be killed in the raid. Then, since, as we have seen, the truth of these two statements is quite compatible with the truth of the statement that, if I do not take precautions, I shall not be killed, the truth of this latter statement cannot be a ground for saying that my taking precautions will not be effective in preventing my death.

Thus, briefly, my method of rebutting the fatalist is to allow him to infer from 'You will not be killed' to 'If you do not take precautions, you will not be killed'; but to point out that, on any sense of 'if' on which this inference is valid, it is impermissible to pass from 'If you do not take precautions, you will not be killed' to 'Your taking precautions will not be effective in preventing your death'. For this to be permissible, the truth of 'If you do not take precautions, you will not be killed' would have to be incompatible with that of 'If you do not take precautions, you will be killed'; but, on the sense of 'if' on which the first step was justified, these would not be incompatible. I prefer to put the matter this way than to make out that there is a sense of 'if' on which these two are indeed incompatible, but on which the first step is unjustified, because it is notoriously difficult to elucidate such a sense of 'if'.

Having arrived at a formulation of the fallacy of the fatalist argument, let us now consider whether the parallel argument to demonstrate the absurdity of attempting to bring about the past is fallacious in the same way. I will abandon the theological example in favour of a magical one. Suppose we come across a tribe who have the following custom. Every second year the young men of the tribe are sent, as part of their initiation ritual, on a lion hunt: they have to prove their manhood. They travel for two days, hunt lions for two days, and spend two days on the return journey; observers go with them, and report to the chief upon their return whether the young men acquitted themselves with bravery or not. The people of the tribe believe that various ceremonies, carried out by the chief, influence the weather, the crops, and so forth. I do not want these ceremonies to be thought of as religious rites, intended to dispose the gods favourably towards them, but simply as performed on the basis of a wholly mistaken system of causal beliefs. While the young men are away from the village the chief performs ceremonies—

dances, let us say—intended to cause the young men to act bravely. We notice that he continues to perform these dances for the whole six days that the party is away, that is to say, for two days during which the events that the dancing is supposed to influence have already taken place. Now there is generally thought to be a *special* absurdity in the idea of affecting the past, much greater than the absurdity of believing that the performance of a dance can influence the behaviour of a man two days' journey away; so we ought to be able to persuade the chief of the absurdity of his continuing to dance after the first four days without questioning his general system of causal beliefs. How are we going to do it?

Since the absurdity in question is alleged to be a *logical* absurdity, it must be capable of being seen to be absurd however things turn out; so I am entitled to suppose that things go as badly for us, who are trying to persuade the chief of this absurdity, as they can do; we ought still to be able to persuade him. We first point out to him that he would not think of continuing to perform the dances after the hunting party has returned; he agrees to that, but replies that that is because at that time he *knows* whether the young men have been brave or not, so there is no longer any point in trying to bring it about that they have been. It is irrelevant, he says, that during the last two days of the dancing they have already either been brave or cowardly: there is still a point in his trying to make them have been brave, because he does not yet know which they have been. We then say that it can be only the first four days of the dancing which could possibly affect the young men's performance; but he replies that experience is against that. There was for several years a chief who thought as we did, and danced for the first four days only; the results were disastrous. On two other occasions, he himself fell ill after four days of dancing and was unable to continue, and again, when the hunting party returned, it proved that the young men had behaved ignobly.

The brief digression into fatalism was occasioned by our noticing that the standard argument against attempting to affect the past was a precise analogue of the standard fatalist argument against attempting to affect the future. Having diagnosed the fallacy in the fatalist argument, my announced intention was to discover whether there was not a similar fallacy in the standard argument against affecting the past. And it indeed appears to me that there is. We say to the chief, 'Why go on dancing now? Either the young men have already been brave, or they have already been cowardly. If they have been brave, then they have been brave whether you dance or not. If they have been cowardly, then they have been cowardly whether you dance or not. If they have been brave, then your dancing now will not be effective in making them have been brave, since they have been brave even if you do not dance. And if they have not been brave, then your dancing will certainly not be effective.

Thus your continuing to dance will in the one case be superfluous, and in the other fruitless: in neither case is there any point in your continuing to dance.' The chief can reply in exactly the way in which we replied to the fatalist. He can say, 'If they have been brave, then indeed there is a sense in which it will be true to say that, even if I do not dance, they will have been brave; but this is not incompatible with its also being true to say that, if I do not dance, they will not have been brave. Now what saying that my continuing to dance is effective in causing them to have been brave amounts to is that it is true both that, if I go on dancing, they have been brave, and that, if I do not dance, they have not been brave. I have excellent empirical grounds for believing both these two statements to be true; and neither is incompatible with the truth of the statement that, if I do not dance, they have been brave, although, indeed, I have no reason for believing *that* statement. Hence, you have not shown that, from the mere hypothesis that they have been brave, it follows that the dancing I am going to do will not be effective in making them have been brave; on the contrary, it may well be that, although they have been brave, they have been brave just *because* I am going to go on dancing; that, if I were not going to go on dancing, they would not have been brave.' This reply sounds sophistical; but it cannot be sophistical if our answer to the fatalist was correct, because it is the exact analogue of that answer.

We now try the following argument: 'Your *knowledge* of whether the young men have been brave or not may affect whether you *think* there is any point in performing the dances; but it cannot really make any difference to the *effect* the dances have on what has happened. If the dances are capable of bringing it about that the young men have acted bravely, then they ought to be able to do that even after you have learned that the young men have *not* acted bravely. But that is absurd, for that would mean that the dances can change the past. But if the dances cannot have any effect after you have learned whether the young men have been brave or not, they cannot have any effect before, either; for the mere state of your knowledge cannot make any difference to their efficacy.' Now since the causal beliefs of this tribe are so different from our own, I could imagine that the chief might simply deny this: he might say that what had an effect on the young men's behaviour was not merely the performance of the dances by the chief as such, but rather their performance by the chief when in a state of ignorance as to the outcome of the hunt. And if he says this, I think there is really no way of dissuading him, short of attacking his whole system of causal beliefs. But I will not allow him to say this, because it would make his causal beliefs so different in kind from ours that there would be no moral to draw for our own case. Before going on to consider his reaction to this argument, however, let us first pause to review the situation.

Suppose, then, that he agrees to our suggestion: agrees, that is, that it is his dancing as such that he wants to consider as bringing about the young men's bravery, and not his dancing in ignorance of whether they were brave. If this is his belief, then we may reasonably challenge him to try dancing on some occasion when the hunting party has returned and the observers have reported that the young men have *not* been brave. Here at last we appear to have hit on something which has no parallel in the case of affecting the future. If someone believes that a certain kind of action is effective in bringing about a subsequent event, I may challenge him to try it out in all possible circumstances: but I cannot demand that he try it out on some occasion when the event is *not* going to take place, since he cannot identify any such occasion independently of his intention to perform the action. Our knowledge of the future is of two kinds: prediction based on causal laws and knowledge in intention. If I think I can predict the non-occurrence of an event, then I cannot consistently also believe that I can do anything to bring it about; that is, I cannot have good grounds for believing, of any action, both that it is in my power to do it, and that it is a condition of the event's occurring. On the other hand, I cannot be asked to perform the action on some occasion when I believe that the event will not take place, when this knowledge lies in my intention to prevent it taking place; for as soon as I accede to the request, I thereby abandon my intention. It would, indeed, be different if we had foreknowledge: someone who thought, like Russell and Ayer, that it is a merely contingent fact that we have memory but not foreknowledge would conclude that the difference I have pointed to does not reveal a genuine asymmetry between past and future, but merely reflects this contingent fact.

If the chief accepts the challenge, and dances when he knows that the young men have not been brave, it seems that he must concede that his dancing does not *ensure* their bravery. There is one other possibility favourable to us. Suppose that he accepts the challenge, but when he comes to try to dance, he unaccountably cannot do so: his limbs simply will not respond. Then we may say, 'It is not your dancing (after the event) which causes them to have been brave, but rather their bravery which makes possible your dancing: your dancing is not, as you thought, an action which it is in your power to do or not to do as you choose. So you ought not to say that you dance in the last two days in order to make them have been brave, but that you try to see whether you can dance, in order to find out whether they have been brave.'

It may seem that this is conclusive; for are not these the only two possibilities? Either he does dance, in which case the dancing is proved not to be a sufficient condition of the previous bravery; or he does not, in which case the bravery must be thought a causal condition of the dancing rather than vice versa. But in fact the situation is not quite so simple.

For one thing, it is not justifiable to demand that the chief should either consider his dancing to be a sufficient condition of the young men's bravery, or regard it as wholly unconnected. It is enough, in order to provide him with a motive for performing the dances, that he should have grounds to believe that there is a significant positive correlation between his dancing and previous brave actions on the part of the young men; so the occurrence of a certain proportion of occasions on which the dancing is performed, although the young men were not brave, is not a sufficient basis to condemn him as irrational if he continues to dance during the last two days. Secondly, while his being afflicted with an otherwise totally inexplicable inability to dance may strongly suggest that the cowardice of the young men renders him unable to dance, and that therefore dancing is not an action which it is in his power to perform as he chooses, any failure to dance that is explicable without reference to the outcome of the hunt has much less tendency to suggest this. Let us suppose that we issue our challenge, and he accepts it. On the first occasion when the observers return and report cowardly behaviour on the part of the young men, he performs his dance. This weakens his belief in the efficacy of the dancing, but does not disturb him unduly; there have been occasions before when the dancing has not worked, and he simply classes this as one of them. On the second occasion when the experiment can be tried, he agrees to attempt it, but, a few hours before the experiment is due to be carried out, he learns that a neighbouring tribe is marching to attack his, so the experiment has to be abandoned; on the third occasion, he is bitten by a snake, and so is incapacitated for dancing. Someone might wish to say, 'The cowardice of the young men caused those events to happen and so prevent the chief from dancing', but such a description is far from mandatory: the chief may simply say that these events were accidental, and in no way *brought about* by the cowardice of the young men. It is true that if the chief is willing to attempt the experiment a large number of times, and events of this kind repeatedly occur, it will no longer appear reasonable to dismiss them as a series of coincidences. If accidents which prevent his dancing occur on occasions when the young men are known to have been cowardly with much greater frequency than, say, in a control group of dancing attempts, when the young men are known to have been brave, or when it is not known how they behaved, then this frequency becomes something that must itself be explained, even though each particular such event already has its explanation.

Suppose now, however, that the following occurs. We ask the chief to perform the dances on some occasion when the hunting party has returned and the observers have reported that the young men have not acquitted themselves with bravery. He does so, and we claim another weakening of his belief that the dancing is correlated with preceding bravery. But later it turns out that, for some reason or other, the observers were lying (say they had been bribed by someone): so after all this is not a counter-example to the law. So we have a third possible outcome. The situation now is this. We challenge the chief to perform the dances whenever he knows that the young men have not been brave, and he accepts the challenge. There are three kinds of outcome: (i) he simply performs the dances; (ii) he is prevented from performing the dances by some occurrence which has a quite natural explanation totally independent of the behaviour of the young men; and (iii) he performs the dances, but subsequently discovers that this was not really an occasion on which the young men had not been brave. We may imagine that he carries out the experiment repeatedly, and that the outcome always falls into one of these three classes; and that outcomes of class (i) are sufficiently infrequent not to destroy his belief that there is a significant correlation between the dancing and the young men's bravery, and outcomes of class (ii) sufficiently infrequent not to make him say that the young men's cowardice renders him incapable of performing the dances. Thus our experiment has failed.

On the other hand, it has not left everything as before. I have exploited the fact that it is frequently possible to discover that one had been mistaken in some belief about the past. I will not here raise the question whether it is *always* possible to discover this, or whether there are beliefs about the past about which we can be *certain* in the sense that nothing could happen to show the belief to have been mistaken. Now before we challenged the chief to perform this series of experiments, his situation was as follows. He was prepared to perform the dancing in order to bring it about that the young men had been brave, but only when he had no information about whether they had been brave or not. The rationale of his doing so was simply this: experience shows that there is a positive correlation between the dancing and the young men's bravery; hence the fact that the dances are being performed makes it more probable that the young men have been brave. But the dancing is something that is in my power to do if I choose: experience does not lead me to recognize it as a possibility that I should try to perform the dances and fail. Hence it is in my power to do something, the doing of which will make it more probable that the young men have been brave: I have therefore every motive to do it. Once he had information, provided by the observers, about the behaviour of the young men, then, under the old dispensation, his attitude changed: he no longer had a motive to perform the dances. We do not have to assume that he was unaware of the possibility that the observers were lying or had made a mistake. It may just have been that he reckoned the probability that they were telling the truth as so high that the performance of the dances after they had made their report would make no significant difference to the probability that the young men had been brave. If they reported the young men as having been brave, there was so little chance of their being wrong that it was not worth while to attempt to diminish this chance by performing the dances; if they reported that the young men had been cowardly, then even the performance of the dances would still leave it overwhelmingly probable that they *had* been cowardly. That is to say, until the series of experiments was performed, the chief was prepared to discount completely the probability conferred by his dancing on the proposition that the young men had been brave in the face of a source of information as to the truth of this proposition of the kind we ordinarily rely upon in deciding the truth or falsity of statements about the past. And the reason for this attitude is very clear: for the proposition that there was a positive correlation between the dancing and the previous bravery of the young men could have been established in the first place only by relying on our ordinary sources of information as to whether the young men had been brave or not.

But if we are to suppose that the series of experiments works out in such a way as not to force the chief to abandon his belief both that there is such a positive correlation and that the dancing is something which it is in his power to do when he chooses, we must suppose that it fairly frequently happens that the observers are subsequently proved to have been making false statements. And I think it is clear that in the process the attitude of the chief to the relative degree of probability conferred on the statement that the young men have been brave by (i) the reports of the observers and (ii) his performance of the dances will alter. Since it so frequently happens that, when he performs the dances *after* having received an adverse report from the observers, the observers prove to have been misreporting, he will cease to think it pointless to perform the dances after having received such an adverse report: he will thus cease to think that he can decide whether to trust the reports of the observers independently of whether he is going to perform the dances or not. In fact, it seems likely that he will come to think of the performance of the dances as itself a ground for distrusting, or even for denying outright, the adverse reports of the observers, even in the absence of any *other* reason (such as the discovery of their having been bribed, or the reports of some other witness) for believing them not to be telling the truth.

The chief began with two beliefs: (i) that there was a positive correlation between his dancing and the previous brave behaviour of the young men; and (ii) that the dancing was something in his power to do as he chose. We are tempted to think of these two beliefs as incompatible, and I described people attempting to devise a series of experiments to convince the chief of this. I tried to show, however, that these experiments could turn out in such a way as to allow the chief to maintain both beliefs. But in the process a third belief, which we naturally take for granted, has had to be abandoned in order to hang on to the first two: the belief, namely, that it is possible for me to find out what has happened (whether the young men have been brave or not) independently of my intentions. The chief no longer thinks that there is any evidence as to whether the young men had been brave or not, the strength of which is unaffected by whether he intends subsequently to perform the dances. And now it appears that there really is a form of incompatibility among these *three* beliefs, in the sense that it is always possible to carry out a series of actions which will necessarily lead to the abandonment of at least one of them. Here there is an exact parallel with the case of affecting the future. We *never* combine the beliefs (i) that an action *A* is positively correlated with the subsequent occurrence of an event *B*; (ii) that the action *A* is in my power to perform or not as I choose; and (iii) that I can know whether *B* is going to take place or not independently of my intention to perform or not to perform the action *A*. The difference between past and future lies in this: that we think that, of any past event, it is in principle possible for me to know whether or not it took place independently of my present intentions; whereas, for many types of future event, we should admit that we are never going to be in a position to have such knowledge independently of our intentions. (If we had foreknowledge, this might be different.) If we insist on hanging on to this belief, for all types of past event, then we cannot combine the two beliefs that are required to make sense of doing something in order that some event should have previously taken place; but I do not know any reason why, if things were to turn out differently from the way they do now, we *could* not reasonably abandon the first of these beliefs rather than either of the other two.

My conclusion therefore is this. If anyone were to claim, of some type of action *A*, (i) that experience gave grounds for holding the performance of *A* as increasing the probability of the previous occurrence of a type of event *E*; and (ii) that experience gave no grounds for regarding *A* as an action which it was ever not in his power to perform—that is, for entertaining the possibility of his trying to perform it and failing—then we could either force him to abandon one or other of these beliefs, or else to abandon the belief (iii) that it was ever possible for him to have knowledge, independent of his intention to perform *A* or not, of whether an event *E* had occurred. Now doubtless most normal human beings would rather abandon either (i) or (ii) than (iii), because we have the prejudice that (iii) must hold good for every type of event: but if someone were, in a particular case, more ready to give up (iii) than (i) or (ii), I cannot see any argument we could use to dissuade him. And so long as he was not dissuaded, he could sensibly speak of performing A in order that E should have occurred. Of course, he could adopt an intermediate position. It is not really necessary, for him to be able to speak of doing A in order that E should have occurred, that he deny all possibility of his trying and failing to perform A. All that is necessary is that he should not regard his being informed, by ordinary means, of the non-occurrence of E as making it more probable that if he tries to perform A, he will fail: for, once he does so regard it, we can claim that he should regard the occurrence of E as making possible the performance of A, in which case his trying to perform A is not a case of trying to bring it about that E has happened, but of finding out whether E has happened. (Much will here depend on whether there is an ordinary causal explanation for the occurrence of E or not.) Now he need not really deny that learning, in the ordinary way, that E has not occurred makes it at all more probable that, if he tries to perform A, he will fail. He may concede that it makes it to some extent more probable, while at the same time maintaining that, even when he has grounds for thinking that E has not occurred, his intention to perform A still makes it more probable than it would otherwise be that E has in fact occurred. The attitude of such a man seems paradoxical and unnatural to us, but I cannot see any rational considerations which would force him out of this position. At least, if there are any, it would be interesting to know what they are: I think that none of the considerations I have mentioned in this paper could serve this purpose.

My theological example thus proves to have been a bad—that is, untypical—example in a way we did not suspect at the time, for it will never lead to a discounting of our ordinary methods of finding out about the past. I may pray that the announcer has made a mistake in not including my son's name on the list of survivors; but once I am convinced that no mistake has been made, I will not go on praying for him to have survived. I should regard this kind of prayer as something to which it was possible to have recourse only when an ordinary doubt about what had happened could be entertained. But just because this example is untypical in this way, it involves no tampering with our ordinary conceptual apparatus at all: this is why it is such a natural thing to do. On my view, then, orthodox Jewish theology is mistaken on this point.

I do not know whether it could be held that part of what people have meant when they have said 'You cannot change the past' is that, for every type of event, it is in principle possible to know whether or not it has happened, independently of one's own intentions. If so, this is not the mere tautology it appears to be, but it does indeed single out what it is that makes us think it impossible to bring about the past.

VIII

THE PARADOXES OF TIME TRAVEL

DAVID LEWIS

Time travel, I maintain, is possible. The paradoxes of time travel are oddities, not impossibilities. They prove only this much, which few would have doubted: that a possible world where time travel took place would be a most strange world, different in fundamental ways from the world we think is ours.

I shall be concerned here with the sort of time travel that is recounted in science fiction. Not all science fiction writers are clear-headed, to be sure, and inconsistent time travel stories have often been written. But some writers have thought the problems through with great care, and their stories are perfectly consistent.[1]

If I can defend the consistency of some science fiction stories of time travel, then I suppose parallel defences might be given of some controversial physical hypotheses, such as the hypothesis that time is circular or the hypothesis that there are particles that travel faster than light. But I shall not explore these parallels here.

What is time travel? Inevitably, it involves a discrepancy between time and time. Any traveler departs and then arrives at his destination; the time elapsed from departure to arrival (positive, or perhaps zero) is the duration of the journey. But if he is a time traveler, the separation in time between departure and arrival does not equal the duration of his journey. He departs; he travels for an hour, let us say; then he arrives. The time he reaches is not the time one hour after his departure. It is later, if he has traveled toward the future; earlier, if he has traveled toward the past. If he has traveled far toward the past, it is

David Lewis, "The Paradoxes of Time Travel". First published in the *American Philosophical Quarterly*, 13 (1976): 145–52. Reprinted by permission of the author and the editor.

The present paper summarizes a series of lectures of the same title, given as the Gavin David Young Lectures in Philosophy at the University of Adelaide in July, 1971. I thank the Australian–American Educational Foundation and the American Council of Learned Societies for research support. I am grateful to many friends for comments on earlier versions of this paper; especially Philip Kitcher, William Newton-Smith, J. J. C. Smart, and Donald Williams. The text as reprinted here includes some small corrections.

[1] I have particularly in mind two of the time travel stories of Robert A. Heinlein: "By his Bootstraps", in Robert A. Heinlein, *The Menace from Earth* (Hicksville, NY: Gnome Press, 1959), and "—All You Zombies—", in Robert A. Heinlein, *The Unpleasant Profession of Jonathan Hoag* (Hicksville, NY: Gnome Press, 1959).

earlier even than his departure. How can it be that the same two events, his departure and his arrival, are separated by two unequal amounts of time?

It is tempting to reply that there must be two independent time dimensions; that for time travel to be possible, time must be not a line but a plane.[2] Then a pair of events may have two unequal separations if they are separated more in one of the time dimensions than in the other. The lives of common people occupy straight diagonal lines across the plane of time, sloping at a rate of exactly one hour of time$_1$ per hour of time$_2$. The life of the time traveler occupies a bent path, of varying slope.

On closer inspection, however, this account seems not to give us time travel as we know it from the stories. When the traveler revisits the days of his childhood, will his playmates be there to meet him? No; he has not reached the part of the plane of time where they are. He is no longer separated from them along one of the two dimensions of time, but he is still separated from them along the other. I do not say that two-dimensional time is impossible, or that there is no way to square it with the usual conception of what time travel would be like. Nevertheless, I shall say no more about two-dimensional time. Let us set it aside, and see how time travel is possible even in one-dimensional time.

The world—the time traveler's world, or ours—is a four-dimensional manifold of events. Time is one dimension of the four, like the spatial dimensions except that the prevailing laws of nature discriminate between time and the others—or rather, perhaps, between various timelike dimensions and various spacelike dimensions. (Time remains one-dimensional, since no two timelike dimensions are orthogonal.) Enduring things are timelike streaks: wholes composed of temporal parts, or *stages*, located at various times and places. Change is qualitative difference between different stages— different temporal parts—of some enduring thing, just as a "change" in scenery from east to west is a qualitative difference between the eastern and western spatial parts of the landscape. If this paper should change your mind about the possibility of time travel, there will be a difference of opinion between two different temporal parts of you, the stage that started reading and the subsequent stage that finishes.

If change is qualitative difference between temporal parts of something, then what doesn't have temporal parts can't change. For instance, numbers can't change; nor can the events of any moment of time, since they cannot be subdivided into dissimilar temporal parts. (We have set aside the case of

² Accounts of time travel in two-dimensional time are found in Jack W. Meiland, "A Two-Dimensional Passage Model of Time for Time Travel", *Philosophical Studies*, 26 (1974): 153–73; and in the initial chapters of Isaac Asimov, *The End of Eternity* (Garden City, NY: Doubleday, 1955). Asimov's denouement, however, seems to require some different conception of time travel.

two-dimensional time, and hence the possibility that an event might be momentary along one time dimension but divisible along the other.) It is essential to distinguish change from "Cambridge change", which can befall anything. Even a number can "change" from being to not being the rate of exchange between pounds and dollars. Even a momentary event can "change" from being a year ago to being a year and a day ago, or from being forgotten to being remembered. But these are not genuine changes. Not just any old reversal in truth-value of a time-sensitive sentence about something makes a change in the thing itself.

A time traveler, like anyone else, is a streak through the manifold of space-time, a whole composed of stages located at various times and places. But he is not a streak like other streaks. If he travels toward the past he is a zig-zag streak, doubling back on himself. If he travels toward the future, he is a stretched-out streak. And if he travels either way instantaneously, so that there are no intermediate stages between the stage that departs and the stage that arrives and his journey has zero duration, then he is a broken streak.

I asked how it could be that the same two events were separated by two unequal amounts of time, and I set aside the reply that time might have two independent dimensions. Instead I reply by distinguishing time itself, *external time* as I shall also call it, from the *personal time* of a particular time traveler: roughly, that which is measured by his wristwatch. His journey takes an hour of his personal time, let us say; his wristwatch reads an hour later at arrival than at departure. But the arrival is more than an hour after the departure in external time, if he travels toward the future; or the arrival is before the departure in external time (or less than an hour after), if he travels toward the past.

That is only rough. I do not wish to define personal time operationally, making wristwatches infallible by definition. That which is measured by my own wristwatch often disagrees with external time, yet I am no time traveler; what my misregulated wristwatch measures is neither time itself nor my personal time. Instead of an operational definition, we need a functional definition of personal time: it is that which occupies a certain role in the pattern of events that comprise the time traveler's life. If you take the stages of a common person, they manifest certain regularities with respect to external time. Properties change continuously as you go along, for the most part, and in familiar ways. First come infantile stages. Last come senile ones. Memories accumulate. Food digests. Hair grows. Wristwatch hands move. If you take the stages of a time traveler instead, they do not manifest the common regularities with respect to external time. But there is one way to assign coordinates to the time traveler's stages, and one way only (apart from the arbitrary choice of a zero point), so that the regularities that hold with respect to this assignment match those that commonly hold with respect to external time. With respect to the correct assignment properties change continuously as you go along, for the most part, and in familiar ways. First come infantile stages. Last come senile ones. Memories accumulate. Food digests. Hair grows. Wristwatch hands move. The assignment of coordinates that yields this match is the time traveler's personal time. It isn't really time, but it plays the role in his life that time plays in the life of a common person. It's enough like time so that we can—with due caution—transplant our temporal vocabulary to it in discussing his affairs. We can say without contradiction, as the time traveler prepares to set out, "Soon he will be in the past". We mean that a stage of him is slightly later in his personal time, but much earlier in external time, than the stage of him that is present as we say the sentence.

We may assign locations in the time traveler's personal time not only to his stages themselves but also to the events that go on around him. Soon Caesar will die, long ago; that is, a stage slightly later in the time traveler's personal time than his present stage, but long ago in external time, is simultaneous with Caesar's death. We could even extend the assignment of personal time to events that are not part of the time traveler's life, and not simultaneous with any of his stages. If his funeral in ancient Egypt is separated from his death by three days of external time and his death is separated from his birth by three score years and ten of his personal time, then we may add the two intervals and say that his funeral follows his birth by three score years and ten and three days of *extended personal time*. Likewise a bystander might truly say, three years after the last departure of another famous time traveler, that "he may even now—if I may use the phrase—be wandering on some plesiosaurus-haunted oolitic coral reef, or beside the lonely saline seas of the Triassic Age".[3] If the time traveler does wander on an oolitic coral reef three years after his departure in his personal time, then it is no mistake to say with respect to his extended personal time that the wandering is taking place "even now".

We may liken intervals of external time to distances as the crow flies, and intervals of personal time to distances along a winding path. The time traveler's life is like a mountain railway. The place two miles due east of here may also be nine miles down the line, in the westbound direction. Clearly we are not dealing here with two independent dimensions. Just as distance along the railway is not a fourth spatial dimension, so a time traveler's

---

[3] H. G. Wells, *The Time Machine, an Invention* (London: Heinemann, 1895), epilogue. The passage is criticized as contradictory in Donald C. Williams, "The Myth of Passage", *Journal of Philosophy*, 48 (1951): 457–72, at 463.

personal time is not a second dimension of time. How far down the line some place is depends on its location in three-dimensional space, and likewise the locations of events in personal time depend on their locations in one-dimensional external time.

Five miles down the line from here is a place where the line goes under a trestle; two miles further is a place where the line goes over a trestle; these places are one and the same. The trestle by which the line crosses over itself has two different locations along the line, five miles down from here and also seven. In the same way, an event in a time traveler's life may have more than one location in his personal time. If he doubles back toward the past, but not too far, he may be able to talk to himself. The conversation involves two of his stages, separated in his personal time but simultaneous in external time. The location of the conversation in personal time should be the location of the stage involved in it. But there are two such stages; to share the locations of both, the conversation must be assigned two different locations in personal time.

The more we extend the assignment of personal time outwards from the time traveler's stages to the surrounding events, the more will such events acquire multiple locations. It may happen also, as we have already seen, that events that are not simultaneous in external time will be assigned the same location in personal time—or rather, that at least one of the locations of one will be the same as at least one of the locations of the other. So extension must not be carried too far, lest the location of events in extended personal time lose its utility as a means of keeping track of their roles in the time traveler's history.

A time traveler who talks to himself, on the telephone perhaps, looks for all the world like two different people talking to each other. It isn't quite right to say that the whole of him is in two places at once, since neither of the two stages involved in the conversation is the whole of him, or even the whole of the part of him that is located at the (external) time of the conversation. What's true is that he, unlike the rest of us, has two different complete stages located at the same time at different places. What reason have I, then, to regard him as one person and not two? What unites his stages, including the simultaneous ones, into a single person? The problem of personal identity is especially acute if he is the sort of time traveler whose journeys are instantaneous, a broken streak consisting of several unconnected segments. Then the natural way to regard him as more than one person is to take each segment as a different person. No one of them is a time traveler, and the peculiarity of the situation comes to this: all but one of these several people vanish into thin air, all but another one appear out of thin air, and there are remarkable resemblances between one at his appearance and another at his vanishing. Why isn't that at least as good a description as the one I gave, on which the several segments are all parts of one time traveler?

I answer that what unites the stages (or segments) of a time traveler is the same sort of mental, or mostly mental, continuity and connectedness that unites anyone else. The only difference is that whereas a common person is connected and continuous with respect to external time, the time traveler is connected and continuous only with respect to his own personal time. Taking the stages in order, mental (and bodily) change is mostly gradual rather than sudden, and at no point is there sudden change in too many different respects all at once. (We can include position in external time among the respects we keep track of, if we like. It may change discontinuously with respect to personal time if not too much else changes discontinuously along with it.) Moreover, there is not too much change altogether. Plenty of traits and traces last a lifetime. Finally, the connectedness and the continuity are not accidental. They are explicable; and further, they are explained by the fact that the properties of each stage depend causally on those of the stages just before in personal time, the dependence being such as tends to keep things the same.[4]

To see the purpose of my final requirement of causal continuity, let us see how it excludes a case of counterfeit time travel. Fred was created out of thin air, as if in the midst of life; he lived a while, then died. He was created by a demon, and the demon had chosen at random what Fred was to be like at the moment of his creation. Much later someone else, Sam, came to resemble Fred as he was when first created. At the very moment when the resemblance became perfect, the demon destroyed Sam. Fred and Sam together are very much like a single person: a time traveler whose personal time starts at Sam's birth, goes on to Sam's destruction and Fred's creation, and goes on from there to Fred's death. Taken in this order, the stages of Fred-*cum*-Sam have the proper connectedness and continuity. But they lack causal continuity, so Fred-*cum*-Sam is not one person and not a time traveler. Perhaps it was pure coincidence that Fred at his creation and Sam at his destruction were exactly alike; then the connectedness and continuity of Fred-*cum*-Sam across the crucial point are accidental. Perhaps instead the demon remembered what Fred was like, guided Sam toward perfect resemblance, watched his progress, and destroyed him at the right moment. Then the connectedness and continuity of Fred-*cum*-Sam have a causal explanation, but of the wrong sort. Either way, Fred's first stages do not depend causally for their properties on Sam's last stages. So the case of Fred and Sam is rightly disqualified as a case of personal identity and as a case of time travel.

---

[4] I discuss the relation between personal identity and mental connectedness and continuity at greater length in "Survival and Identity", in Amélie Rorty (ed), *The Identities of Persons* (Berkeley, Calif : University of California Press, 1976)

We might expect that when a time traveler visits the past there will be reversals of causation. You may punch his face before he leaves, causing his eye to blacken centuries ago. Indeed, travel into the past necessarily involves reversed causation. For time travel requires personal identity—he who arrives must be the same person who departed. That requires causal continuity, in which causation runs from earlier to later stages in the order of personal time. But the orders of personal and external time disagree at some point, and there we have causation that runs from later to earlier stages in the order of external time. Elsewhere I have given an analysis of causation in terms of chains of counterfactual dependence, and I took care that my analysis would not rule out causal reversal *a priori*.[5] I think I can argue (but not here) that under my analysis the direction of counterfactual dependence and causation is governed by the direction of other *de facto* asymmetries of time. If so, then reversed causation and time travel are not excluded altogether, but can occur only where there are local exceptions to these asymmetries. As I said at the outset, the time traveler's world would be a most strange one.

Stranger still, if there are local—but only local—causal reversals, then there may also be causal loops: closed causal chains in which some of the causal links are normal in direction and others are reversed. (Perhaps there must be loops if there is reversal; I am not sure.) Each event on the loop has a causal explanation, being caused by events elsewhere on the loop. That is not to say that the loop as a whole is caused or explicable. It may not be. Its inexplicability is especially remarkable if it is made up of the sort of causal processes that transmit information. Recall the time traveler who talked to himself. He talked to himself about time travel, and in the course of the conversation his older self told his younger self how to build a time machine. That information was available in no other way. His older self knew how because his younger self had been told and the information had been preserved by the causal processes that constitute recording, storage, and retrieval of memory traces. His younger self knew, after the conversation, because his older self had known and the information had been preserved by the causal processes that constitute telling. But where did the information come from in the first place? Why did the whole affair happen? There is simply no answer. The parts of the loop are explicable, the whole of it is not. Strange! But not impossible, and not too different from inexplicabilities we are already inured to. Almost everyone agrees that God, or the Big Bang, or the entire infinite past of the universe, or the decay of a tritium atom, is uncaused and inexplicable. Then if these are possible, why not also the inexplicable causal loops that arise in time travel?

---

[5] "Causation", *Journal of Philosophy*, 70 (1973): 556–67; the analysis relies on the analysis of counterfactuals given in my *Counterfactuals* (Oxford: Blackwell, 1973).

I have committed a circularity in order not to talk about too much at once, and this is a good place to set it right. In explaining personal time, I presupposed that we were entitled to regard certain stages as comprising a single person. Then in explaining what united the stages into a single person, I presupposed that we were given a personal time order for them. The proper way to proceed is to define personhood and personal time simultaneously, as follows. Suppose given a pair of an aggregate of person-stages, regarded as a candidate for personhood, and an assignment of coordinates to those stages, regarded as a candidate for his personal time. Iff the stages satisfy the conditions given in my circular explanation with respect to the assignment of coordinates, then both candidates succeed: the stages do comprise a person and the assignment is his personal time.

I have argued so far that what goes on in a time travel story may be a possible pattern of events in four-dimensional space-time with no extra time dimension; that it may be correct to regard the scattered stages of the alleged time traveler as comprising a single person; and that we may legitimately assign to those stages and their surroundings a personal time order that disagrees sometimes with their order in external time. Some might concede all this, but protest that the impossibility of time travel is revealed after all when we ask not what the time traveler *does*, but what he *could do*. Could a time traveler change the past? It seems not: the events of a past moment could no more change than numbers could. Yet it seems that he would be as able as anyone to do things that would change the past if he did them. If a time traveler visiting the past both could and couldn't do something that would change it, then there cannot possibly be such a time traveler.

Consider Tim. He detests his grandfather, whose success in the munitions trade built the family fortune that paid for Tim's time machine. Tim would like nothing so much as to kill Grandfather, but alas he is too late. Grandfather died in his bed in 1957, while Tim was a young boy. But when Tim has built his time machine and traveled to 1920, suddenly he realizes that he is not too late after all. He buys a rifle; he spends long hours in target practice; he shadows Grandfather to learn the route of his daily walk to the munitions works; he rents a room along the route; and there he lurks, one winter day in 1921, rifle loaded, hate in his heart, as Grandfather walks closer, closer, ...

Tim can kill Grandfather. He has what it takes. Conditions are perfect in every way: the best rifle money could buy, Grandfather an easy target only twenty yards away, not a breeze, door securely locked against intruders, Tim a good shot to begin with and now at the peak of training, and so on. What's to stop him? The forces of logic will not stay his hand! No powerful chaperone stands by to defend the past from interference. (To imagine such a chaperone, as some authors do, is a boring evasion, not needed to make Tim's story consistent.) In short, Tim is as much able to kill Grandfather as anyone ever is to kill anyone. Suppose that down the street another sniper, Tom, lurks waiting for another victim, Grandfather's partner. Tom is not a time traveler, but otherwise he is just like Tim: same make of rifle, same murderous intent, same everything. We can even suppose that Tom, like Tim, believes himself to be a time traveler. Someone has gone to a lot of trouble to deceive Tom into thinking so. There's no doubt that Tom can kill his victim; and Tim has everything going for him that Tom does. By any ordinary standards of ability, Tim can kill Grandfather.

Tim cannot kill Grandfather. Grandfather lived, so to kill him would be to change the past. But the events of a past moment are not subdivisible into temporal parts and therefore cannot change. Either the events of 1921 time-lessly do include Tim's killing of Grandfather, or else they timelessly don't. We may be tempted to speak of the "original" 1921 that lies in Tim's personal past, many years before his birth, in which Grandfather lived; and of the "new" 1921 in which Tim now finds himself waiting in ambush to kill Grandfather. But if we do speak so, we merely confer two names on one thing. The events of 1921 are doubly located in Tim's (extended) personal time, like the trestle on the railway, but the "original" 1921 and the "new" 1921 are one and the same. If Tim did not kill Grandfather in the "original" 1921, then if he does kill Grandfather in the "new" 1921, he must both kill and not kill Grandfather in 1921—in the one and only 1921, which is both the "new" and the "original" 1921. It is logically impossible that Tim should change the past by killing Grandfather in 1921. So Tim cannot kill Grand-father.

Not that past moments are special; no more can anyone change the present or the future. Present and future momentary events no more have temporal parts than past ones do. You cannot change a present or future event from what it was originally to what it is after you change it. What you *can* do is to change the present or the future from the unactualized way they would have been without some action of yours to the way they actually are. But that is not an actual change: not a difference between two successive actualities. And Tim can certainly do as much; he changes the past from the unactualized way it would have been without him to the one and only way it actually is. To "change" the past in this way, Tim need not do anything momentous; it is enough just to be there, however unobtrusively.

You know, of course, roughly how the story of Tim must go on if it is to be consistent: he somehow fails. Since Tim didn't kill Grandfather in the "original" 1921, consistency demands that neither does he kill Grandfather in the "new" 1921. Why not? For some commonplace reason. Perhaps some noise distracts him at the last moment, perhaps he misses despite all his target practice, perhaps his nerve fails, perhaps he even feels a pang of unaccustomed mercy. His failure by no means proves that he was not really able to kill Grandfather. We often try and fail to do what we are able to do. Success at some tasks requires not only ability but also luck, and lack of luck is not a temporary lack of ability. Suppose our other sniper, Tom, fails to kill Grandfather's partner for the same reason, whatever it is, that Tim fails to kill Grandfather. It does not follow that Tom was unable to. No more does it follow in Tim's case that he was unable to do what he did not succeed in doing.

We have this seeming contradiction: "*Tim doesn't but can, because he has what it takes*" versus "*Tim doesn't, and can't, because it's logically impossible to change the past*". I reply that there is no contradiction. Both conclusions are true, and for the reasons given. They are compatible because "can" is equivocal.

To say that something can happen means that its happening is compossible with certain facts. *Which* facts? That is determined, but sometimes not determined well enough, by context. An ape can't speak a human language—say, Finnish—but I can. Facts about the anatomy and operation of the ape's larynx and nervous system are not compossible with his speaking Finnish. The corresponding facts about my larynx and nervous system are compossible with my speaking Finnish. But don't take me along to Helsinki as your interpreter: I can't speak Finnish. My speaking Finnish is compossible with the facts considered so far, but not with further facts about my lack of training. What I can do, relative to one set of facts, I cannot do, relative to another, more inclusive, set. Whenever the context leaves it open which facts are to count as relevant, it is possible to equivocate about whether I can speak Finnish. It is likewise possible to equivocate about whether it is possible for me to speak Finnish, or whether I am able to, or whether I have the ability or capacity or power or potentiality to. Our many words for much the same thing are little help since they do not seem to correspond to different fixed delineations of the relevant facts.

Tim's killing Grandfather that day in 1921 is compossible with a fairly rich set of facts: the facts about his rifle, his skill and training, the unobstructed line of fire, the locked door and the absence of any chaperone to defend the past, and so on. Indeed it is compossible with all the facts of the sorts we would ordinarily count as relevant in saying what someone can do. It is compossible with all the facts corresponding to those we deem relevant in Tom's case. Relative to these facts, Tim can kill Grandfather. But his killing Grandfather is not compossible with another, more inclusive set of facts. There is the simple fact that Grandfather was not killed. Also there are various other facts about Grandfather's doings after 1921 and their effects: Grandfather begat Father in 1922 and Father begat Tim in 1949. Relative to these facts, Tim cannot kill Grandfather. He can and he can't, but under different delineations of the relevant facts. You can reasonably choose the narrower delineation, and say that he can; or the wider delineation, and say that he can't. But choose. What you mustn't do is waver, say in the same breath that he both can and can't, and then claim that this contradiction proves that time travel is impossible.

Exactly the same goes for Tom's parallel failure. For Tom to kill Grandfather's partner also is compossible with all facts of the sorts we ordinarily count as relevant, but not compossible with a larger set including, for instance, the fact that the intended victim lived until 1934. In Tom's case we are not puzzled. We say without hesitation that he can do it, because we see at once that the facts that are not compossible with his success are facts about the future of the time in question and therefore not the sort of facts we count as relevant in saying what Tom can do.

In Tim's case it is harder to keep track of which facts are relevant. We are accustomed to exclude facts about the future of the time in question, but to include some facts about its past. Our standards do not apply unequivocally to the crucial facts in this special case: Tim's failure, Grandfather's survival, and his subsequent doings. If we have foremost in mind that they lie in the external future of that moment in 1921 when Tim is almost ready to shoot, then we exclude them just as we exclude the parallel facts in Tom's case. But if we have foremost in mind that they precede that moment in Tim's extended personal time, then we tend to include them. To make the latter be foremost in your mind, I chose to tell Tim's story in the order of his personal time, rather than in the order of external time. The fact of Grandfather's survival until 1957 had already been told before I got to the part of the story about Tim lurking in ambush to kill him in 1921. We must decide, if we can, whether to treat these personally past and externally future facts as if they were straightforwardly past or as if they were straightforwardly future.

Fatalists—the best of them—are philosophers who take facts we count as irrelevant in saying what someone can do, disguise them somehow as facts of a different sort that we count as relevant, and thereby argue that we can do less than we think—indeed, that there is nothing at all that we don't do but can. I am not going to vote Republican next fall. The fatalist argues that, strange to say, I not only won't but can't; for my voting Republican is not compossible with the fact that it was true already in the year 1548 that I was not going to vote Republican 428 years later. My rejoinder is that this is a fact, sure enough; however, it is an irrelevant fact about the future masquerading as a relevant fact about the past, and so should be left out of account in saying what, in any ordinary sense, I can do. We are unlikely to be fooled by the fatalist's methods of disguise in this case, or other ordinary cases. But in cases of time travel, precognition, or the like, we're on less familiar ground, so it may take less of a disguise to fool us. Also, new methods of disguise are available, thanks to the device of personal time.

Here's another bit of fatalist trickery. Tim, as he lurks, already knows that he will fail. At least he has the wherewithal to know it if he thinks, he knows it implicitly. For he remembers that Grandfather was alive when he was a boy, he knows that those who are killed are thereafter not alive, he knows (let us suppose) that he is a time traveler who has reached the same 1921 that lies in his personal past, and he ought to understand—as we do—why a time traveler cannot change the past. What is known cannot be false. So his success is not only not compossible with facts that belong to the external future and his personal past, but also is not compossible with the present fact of his knowledge that he will fail. I reply that the fact of his foreknowledge, at the moment while he waits to shoot, is not a fact entirely about that moment. It may be divided into two parts. There is the fact that he then believes (perhaps only implicitly) that he will fail; and there is the further fact that his belief is correct, and correct not at all by accident, and hence qualifies as an item of knowledge. It is only the latter fact that is not compossible with his success, but it is only the former that is entirely about the moment in question. In calling Tim's state at that moment knowledge, not just belief, facts about personally earlier but externally later moments were smuggled into consideration.

I have argued that Tim's case and Tom's are alike, except that in Tim's case we are more tempted than usual—and with reason—to opt for a semi-fatalist mode of speech. But perhaps they differ in another way. In Tom's case, we can expect a perfectly consistent answer to the counterfactual question: what if Tom had killed Grandfather's partner? Tim's case is more difficult. If Tim had killed Grandfather, it seems offhand that contradictions would have been true. The killing both would and wouldn't have occurred. No Grandfather, no Father; no Father, no Tim; no Tim, no killing. And for good measure: no Grandfather, no family fortune; no fortune, no time machine; no time machine, no killing. So the supposition that Tim killed Grandfather seems impossible in more than the semi-fatalistic sense already granted.

If you suppose Tim to kill Grandfather and hold all the rest of his story fixed, of course you get a contradiction. But likewise if you suppose Tom to kill Grandfather's partner and hold the rest of his story fixed—including the part that told of his failure—you get a contradiction. If you make *any* counterfactual supposition and hold all else fixed you get a contradiction. The thing to do is rather to make the counterfactual supposition and hold all else as close to fixed as you consistently can. That procedure will yield perfectly consistent answers to the question: what if Tim had killed Grandfather? In that case, some of the story I told would not have been true. Perhaps Tim might have been the time-traveling grandson of someone else. Perhaps he might have been the grandson of a man killed in 1921 and miraculously resurrected. Perhaps he might have been not a time traveler at all, but rather someone created out of nothing in 1920 equipped with false memories of a personal past that never was. It is hard to say what is the least revision of Tim's story to make it true that Tim kills Grandfather, but certainly the contradictory story in which the killing both does and doesn't occur is not the least revision. Hence it is false (according to the unrevised story) that if Tim had killed Grandfather then contradictions would have been true.

What difference would it make if Tim travels in branching time? Suppose that at the possible world of Tim's story the space-time manifold branches; the branches are separated not in time, and not in space, but in some other way. Tim travels not only in time but also from one branch to another. In one branch Tim is absent from the events of 1921; Grandfather lives; Tim is born, grows up, and vanishes in his time machine. The other branch diverges from the first when Tim turns up in 1920; there Tim kills Grandfather and Grandfather leaves no descendants and no fortune; the events of the two branches differ more and more from that time on. Certainly this is a consistent story; it is a story in which Grandfather both is and isn't killed in 1921 (in the different branches); and it is a story in which Tim, by killing Grandfather, succeeds in preventing his own birth (in one of the branches). But it is not a story in which Tim's killing of Grandfather both does occur and doesn't: it simply does, though it is located in one branch and not the other. And it is not a story in which Tim changes the past. 1921 and later years contain the events of both branches, coexisting somehow without interaction. It remains true at all the personal times of Tim's life, even after the killing, that Grandfather lives in one branch and dies in the other.

4

THE TOPOLOGY OF TIME

IX

RELATIONISM AND TEMPORAL TOPOLOGY: PHYSICS OR METAPHYSICS?

ROBIN LE POIDEVIN

1. INTRODUCTION

To what extent arc disputes concerning the nature of time a matter of philosophical, rather than empirical, investigation? The basis of the division, I take it, is partly epistemic: philosophical questions are open to a priori argument, empirical questions (trivially) not. But there is also a modal distinction: a philosophical theory of time would typically hold that time had such and such a property as a matter of *necessity*, whereas empirical investigation typically discovers only contingent properties of time. This rough, and no doubt fallible, guide allows us to classify disputes concerning time on one side of the physics–metaphysics boundary or the other.

Questions such as 'Does time flow?' and 'Could there be time without change?' are widely regarded as philosophical in character. The issues both concern the question whether there is any *incoherence* in the very concept of temporal flow and changeless time respectively. If there is, then empirical results are beside the point. Another paradigmatically philosophical dispute is that between relationism and absolutism, that is, whether times are logical constructions out of events and their relations, or quite independent of their contents. Indeed questions concerning the possibility of reduction of one category of thing into another typify metaphysical enquiry.

When we come to the question 'What constitutes the direction of time?', classification becomes more difficult. Causal theorists of time, who regard temporal relations as reducible to causal (usually: causal connectibility) relations, will naturally regard the question as a philosophical one. Their answer is that the direction of time is constituted by the direction of causation, and

Robin Le Poidevin, 'Relationism and Temporal Topology: Physics or Metaphysics?', *Philosophical Quarterly*, 40 (1990): 419–32. Reprinted with permission. The Postscript to this essay was written specially for this volume. Earlier versions of the essay were presented at St Andrews, Dubrovnik, and Sheffield. I am very grateful to those present for their comments.

it is an a priori truth that the causal relation is asymmetric.[1] However, it is also sometimes asserted that the direction of time is constituted by asymmetries in physical laws, or even by *de facto* asymmetries in physical processes.[2] Where precisely the asymmetry lies here is clearly a matter of empirical investigation.

Intriguingly, there is a whole range of questions, the orthodox view of which has undergone a complete reversal. Examples of such questions are: 'Does time have a beginning and/or end?', 'Is time branching or non-branching?', 'Does time have a linear or a closed structure?', 'Is time dense or discrete?' Questions of this kind concern the topological structure of time, and were once regarded as susceptible only to a priori argument. In the *Physics* Aristotle argues that time is of necessity unbounded and dense.[3] In the *Treatise* Hume argues that (given his particular brand of empiricism) time is of necessity discrete.[4] The Aristotelian attitude lingers on in some modern writers such as Prior, who finds incoherence in the notions of branching and non-unified time,[5] and Swinburne, who similarly objects to the notions of bounded and closed time.[6] But these views are the exception in modern debate. Since Gödel[7] pointed out that the results of General Relativity allowed for non-standard temporal topologies, the philosophical orthodoxy has shifted towards regarding time as having its topological properties only contingently. Grünbaum, for example, in his development of the causal theory of time, explicitly allowed for the possibility that time might be closed rather than linear.[8] And the view that theories concerning the topology of time are empirical has recently been forcefully expressed in William Newton-Smith's book *The Structure of Time*.[9]

[1] See e.g. A. Grünbaum, *Philosophical Problems of Space and Time*, 2nd edn. (Dordrecht: Reidel, 1973). Of course, one may take the arrow of causation to fix the arrow of time without being a causal theorist in the sense defined here. See D. H. Mellor, *Real Time* (Cambridge: Cambridge University Press, 1981) and D. Lewis, 'Counterfactual Dependence and Time's Arrow', *Noûs*, 13 (1979): 455–76.

[2] On lawlike asymmetry, see L. Sklar, *Space, Time, and Spacetime* (Berkeley, Calif.: University of California Press, 1974), 372–8, 379–94. On *de facto* asymmetry, see ibid. 358, 404–11; Bas C. van Fraassen, *An Introduction to Space and Time*, 2nd edn. (New York: Columbia University Press, 1985), ch. 3. 3

[3] For proofs of density, see Aristotle, *Physics*, VI, esp $232^b 20$ f. and $233^b 33$. For the proof of unboundedness, see $251^b 19–23$.

[4] D. Hume, *A Treatise of Human Nature*, ed. L. A. Selby-Bigge, rev. P. H. Nidditch (Oxford: Clarendon Press, 1978), 26–33.

[5] A. Prior, *Past, Present and Future* (Oxford: Clarendon Press, 1967), 75.

[6] R. G. Swinburne, *Space and Time* (London: Macmillan, 1968), 109, 207.

[7] K. Gödel, 'An Example of a New Type of Cosmological Solution of Einstein's Field Equations of Gravitation', *Review of Modern Physics*, 21 (1949), 446–50.

[8] Grünbaum, *Philosophical Problems of Space and Time*, ch. 7.

[9] W. H. Newton-Smith, *The Structure of Time* (London: Routledge & Kegan Paul, 1980), pp. xi, 221, 230.

One of the reasons for this volte-face, although I put this forward simply as a plausible conjecture, may have been the realization that the supposed demonstrations of incoherence in, for example, bounded, non-unified, or closed time are suspect. (Newton-Smith ably dismantles such arguments.[10]) But if we cannot discover an incoherence in such non-standard topologies, then we lack any grounds to suppose that time has its topological properties as a matter of necessity. It is then a short step from this to the view that topology is essentially an empirical concern. I mean by this, not that observation completely determines a topology, but that it is part of a *physical* theory that time has a given topology, and this plays an explanatory role in the interpretation of empirical results. The task for the philosopher, then, is not to show *that* time has such and such a topology, but rather to demonstrate the consequences of supposing it to have such and such a topology. For example, it is an interesting philosophical question as to how our concept of causality needs to be modified in order to take account of closed time.

I follow this current orthodoxy on time's topology, and therefore find it surprising to discover that relationism, which we earlier characterized as a paradigmatically philosophical doctrine, has consequences for time's topology. This is the result I aim to present in this essay. I shall begin in Section 2 by characterizing relationism and explaining why it is plausible to regard it as a purely philosophical theory. In Section 3 of the paper I shall show what consequences relationism has for temporal topology. I shall end by considering two rival conclusions that can be drawn from this result: one is that temporal topology is, after all, a matter for a priori debate; the other is that relationism poaches on empirical preserves.

2. RELATIONISM

In his famous correspondence with Clarke, Leibniz advances the anti-Newtonian thesis that times are logical constructions out of events. As he puts it: 'instants, consider'd without the things, are nothing at all . . . they consist only in the successive order of things'.[11] For Leibniz, one of the chief merits of this view is that it disposes of a puzzle concerning creation, namely, why did God create the universe at the time he did, rather than at an earlier or a later time? Leibniz's answer is that, if times are just successive changes, then necessarily the beginning of the universe coincides with the first moment of

---

[10] Ibid. 61–5 (against Prior on closed time); 77–8 (against Swinburne on closed time); 79–95 (against Swinburne and Quinton on non-unity); 97–9 (against Aristotle on boundedness).
[11] *The Leibniz–Clarke Correspondence*, ed. H. G. Alexander (Manchester: Manchester University Press, 1956), 3rd paper, sect. 6.

time. Since, however, the puzzle depends in large part upon our unwilling-ness to attribute random choices to God, the corresponding secular question 'Why did the universe come into existence at the moment it did?' has rather less bite.

But even stripped of its theological support, the Leibnizian view of time is a plausible one. Part of the motivation for it is epistemological. Not only are events, on the face of it, less mysterious entities than instants, they are clearly things with which we can causally interact. This is important if we are to reconcile the intuition that times can be the objects of thought and reference with the intuition that causal theories of knowledge and reference are at least approximately correct.[12]

However, the modern doctrine known as relationism regards times as constructions, not just out of actual events, but out of actual *and possible* events.[13] To put it another way, whereas for Leibniz times are changes, for the relationist times are *possibilities* of change. We can represent one rela-tionist doctrine symbolically as follows:

  (A) $(\exists t)(R_n(t, e)) \leftrightarrow \Diamond \exists x(\text{Event}(x) \,\&\, R_n(x, e))$.

That is, there exists a time t which is n units before/after some actual event e if, and only if, it is possible that there should exist an event n units before/after e. This now provides us with a reductionist analysis of times:

  (B) a time t, located n units before/after event e is just the collection of actual and possible events located n units before/after e.

(A) sets out the necessary and sufficient conditions for the existence of a time, (B) says what times are. Note the term 'collection' in (B). The relation-ist analysis is sometimes represented as 'a time is the *set* of actual and possible events . . . etc.'. This is not plausible if 'set' refers to an abstract object. In contrast to the set of certain specified physical objects, the *collect-ion* or *aggregate* of those objects shares at least some of the physical characteristics of its members.[14] The reference to an *actual* event in (A) and (B) is crucial. Suppose the relationist restricted himself entirely to purely possible events in his analysis. This would completely disable his construct-ion of the time-series, for he then has no means to distinguish one time from

[12] Bob Hale, however (in *Abstract Objects* (Oxford: Blackwell, 1987)), has argued that causal theories of knowledge which are not implausibly strong may be consistent with a Platonist concep-tion of abstract objects.

[13] 'Relationism' is sometimes the name given to Leibniz's position. The position I call relationism Newton-Smith calls modal reductionism. In this I follow J. N. Butterfield, 'Relationism and Possible Worlds', *British Journal for the Philosophy of Science*, 35 (1984): 101–13.

[14] For further development of this see T. Burge, 'A Theory of Aggregates', *Noûs*, 11 (1977): 97–117

another. It would be entirely arbitrary to link one time with a possible event of type F and another time with a possible event of type G. The only feature of these events which is relevant for the purposes of constructing an ordered series is the temporal relation they stand in to some actual event. A further reason for retaining reference to actual events is this: the relationist will want to distinguish between *actual* times and *merely possible* times. Unless he defines actual times as those standing in some relation to actual events, he has no means of retaining this distinction. (As we shall see, however, he still faces difficulties in retaining it.)

The relationist analysis as we have presented it will have to be modified somewhat if it is not to lead to absurdities. Newton-Smith has pointed out that if we construe the modality in the analysis as simple logical modality, relationism entails that between any two events (or, to take Newton-Smith's example, between any two parts of an event) there is a period of empty time.[15] This is because, however close the events (or parts of the event) are, it is always *logically* possible that an event should occur between them. So the relationist biconditional will automatically generate a period of arbitrary length between the events. This is obviously unacceptable. As pointed out above, the distinction between actual and purely possible times needs to be maintained. The simplest way to deal with this problem, I suggest, is to modify the relationist account as follows: there exists a time, t, n units before/after an actual event, e, if, and only if, it is possible for there to be an event n units before/after e compatible with no disturbance of the actual temporal relations between actual events. There is no implication here that there must be a possible world in which it is the case *both* that there is an event n units . . . etc. *and* that exactly the same events occur in that world as occur in this world. The condition is rather that it must not be the case that any world in which there exists an event n units . . . etc. is also one in which there are some events which stand in a different temporal relationship from the one they stand in in the actual world. We shall come back to the question of how to construe the modality of the analysis in Section 3.

We may ask: what is the motivation for moving from actual to possible events? One source of the motivation arises from considering cases of temporal vacua: periods of time in the absence of change. Hard-line reductionists will deny the possibility of vacua, but one way of accommodating them without abandoning a reductionist position is to regard vacua as possibilities of change. Now here it might be thought that (A) is in tension with (B). (A) guarantees that a temporal vacuum will consist of a number of distinct times.

---

[15] Newton-Smith, *The Structure of Time*, 44. Newton-Smith recommends that we construe the modality as *physical*, thus precluding such absurdities. I explain in the text why I do not think this is a wise choice for the relationist.

But (B) appears to entail that these times collapse into a single moment. For each of the times making up a temporal vacuum will correspond to the same collection of actual and possible events. There are no actual events in a vacuum, and what is logically possible does not change over time. So (B), it appears, implies that vacua are literally instantaneous.

One answer to this is to point out that *type* possibilities do not change over time, but *token* ones do. The point can be made by casting the issue in terms of worlds and possibilia: it is possible for an event of kind F to occur at t if, and only if, there is at least one world in which a token event of kind F occurs at t; and it is possible for an event of kind F to occur at a different time, t^*, if, and only if, there exists at least one world in which a token event of kind F occurs at t^*. Now either we have two distinct token events here, or, if we do not, then we have at least two different world-occurrences of the same event. So the answer to this puzzle is that the relationist should identify times, not with events, but with (possible) world-occurrences of events. Different times may correspond to the same collection of events, but not to the same collection of world-occurrences of events. We should therefore replace (B) with:

> (B′) a time t, located n units before/after event e is just the collection of actual and possible world-occurrences of actual or possible events located n units before/after e.[16]

(B′), unlike (B), does *not* imply that temporal vacua are instantaneous. (A difficulty here is how precisely to construe 'world-occurrences'. One seems to be faced with a dilemma: if two 'world-occurrences' are to be regarded as genuinely distinct things, then they should be regarded as different counterparts (in Lewis's sense[17]) of an event. But we may not wish to be committed to counterpart theory. If, however, we reject counterpart theory to allow genuine transworld identity of objects and events, then two 'world-occurrences' of an event will just be one and the same event. The very large issue this raises is whether relationism, which essentially involves modal concepts, is best served by a realist account of those concepts, for talk of token possibilia makes little sense outside modal realism. It would be ironic if relationism, motivated as it is in part by considerations of ontological economy, found itself committed to a plurality of worlds! But this is a topic for another occasion. Having made these remarks, we will leave (B′) as it is.)

---

[16] I argue for the possibility of vacua in my book *Change, Cause and Contradiction* (London: Macmillan, 1991), ch. 6. See also Sydney Shoemaker, 'Time without Change', *Journal of Philosophy*, 66 (1969): 363–81 [Essay IV in this volume].

[17] See D. Lewis, *On the Plurality of Worlds* (Oxford: Blackwell, 1986).

There is a further reason for preferring relationism over Leibniz's non-modal reductionism, and that is the fact that times can be the subjects of modal and counterfactual statements. Consider the following:

This heatwave might have been over by *Tuesday*.
1914 might not have been the first year of the war.
If the 200th anniversary of the French Revolution had not happened *this summer*, we would have been staying in Paris.

If, as seems plausible, these sentences are genuinely expressive of *de re* modality, then we must understand them by taking the italicized terms to designate rigidly the times they actually refer to. That is, '1914', as used in the above sentence, refers to that very time in all possible worlds where that time exists. And in different possible worlds (to retain the possible world idiom) it may have different properties. Now if we think of times as being just collections of actual events, we can make no sense of sentences like the three examples above. If, as a matter of fact, the assassination of Archduke Ferdinand occurred in 1914, then on the Leibnizian view this event is essentially constitutive of that time. It would then be as incoherent to say that that event might not have taken place in 1914 as it would to say that the assassination of Archduke Ferdinand might not have been the assassination of Archduke Ferdinand. In other words, Leibniz's simple reductionism entails an unacceptable essentialism with regard to times.

It should now be clear why relationism is plausibly regarded as a philosophical doctrine. The motivation for reductionist theories of time is in part epistemological: we should seek to analyse entities such as times in terms of other entities our knowledge of which is relatively unproblematic. And we have seen that the relationist position in particular is a natural one to adopt if we are concerned with the interpretation of modal statements about times. Thus the route we have taken to relationism has been entirely an a priori one. I shall now set about demonstrating its surprising consequences.

3. TOPOLOGY

Prior, Swinburne, and many other writers favour what may be called the 'standard topology' for time: that is, time as boundless, continuous, linear and non-branching.[18] (I shall be defining these topological terms in what follows.) Those who regard topological theories of time as philosophical hold

---

[18] It is also standardly assumed that time is unified and orientable, but I have been unable to see any connection between relationism and these topological properties.

in addition that time has its topological properties as a matter of necessity. The combination of these positions I shall call the traditional view. The traditional view sees non-standard topologies for time as incoherent, while allowing of course that they can be given coherent mathematical definitions. ('There's nothing wrong with boundedness as a property, it's just that it couldn't possibly be a property of *time*,' they say.) What I want to show in this section is how relationism provides support for the traditional view.

Boundless versus Bounded Time

Time is boundless if, and only if, it has neither a beginning nor an end. Relationism entails that time is boundless, and the proof of this goes as follows. Suppose, for *reductio*, that t is the last (or first) moment of time, where t is n units after (before) an event e. It is logically possible that there should exist another event $2n$ units after (before) e. On the relationist analysis, this entails the existence of an actual time n units after (before) t. Consequently, t cannot after all be the last (first) moment of time. The argument applies again to this later (earlier) time. So time cannot have either a beginning or an end. And since we are making no contingent assumptions, the proof also entails that time is boundless as a matter of necessity.

One way of avoiding this consequence is to insist that the possibility of an event standing in such and such a relation to other events or times entails only the *possibility* of there being a time standing in that relation. This move, however, undermines the relationist's treatment of vacua: there can be a period of time without change precisely because the absence of change does not entail the impossibility of change.

Another way of avoiding the topological consequence is to construe the modality in the relationist analysis as *physical*, rather than purely logical. (Newton-Smith argues that this is how we should in any case construe the modality.[19]) Here physical possibility may be defined in terms of compatibility with physical law. However, if the relationist construes the modality like this, then he undermines his way of coping with the possibility of temporal vacua. It was the possibility of such vacua, recall, which motivated the move away from Leibniz's simple reductionist account of times as changes to the relationist account of times as possibilities of change. Now all temporal vacua allow the logical possibility of change, but not all vacua allow the *physical possibility* of change, that is, at least some vacua may arise in circumstances which make it physically impossible for there to be events

---

[19] See n. 15.

during the period of the vacuum.[20] And if the world is deterministic, then *no* vacuum is compatible with the possibility of change. A further motivation for construing the modality as purely logical rather than physical comes from a direction already mentioned: our modal judgements about times. Times, we remarked, do not have (all) their properties essentially. Some events which did occur at those times need not have done. And some events which did not in fact occur at them might have done. Now we need to construe the 'might have' as widely as possible to take into account statements like: 'It might have been the case that all bodies lost their gravitational pull on other bodies yesterday', and other modal statements concerned with physically impossible events. If times were just *physically* possible world-occurrences of events, then they could not be the subject of true modal statements which nevertheless conflict with physical law. Times, as we might put it, have wider modal extension than laws. So the relationist has good reason to resist a narrow construal of the modality in (A) and (B′).

Continuous versus Discontinuous Time

If, for any two instants, there is a third existing between them, then time is either dense or continuous. Time is discrete if, and only if, for any instant, there is a unique subsequent (and a unique antecedent) instant. The difference between density and continuity is this: a dense ordering is one which is isomorphic to the set of rational numbers, a continuous ordering is one which is isomorphic to the set of real numbers. Relationism entails that time is continuous, and the proof of this goes as follows. Suppose that there exist two times, t and $t′$, separated by an interval of $2n$ units. It is logically possible that there should exist an event, n units after t, where n ranges over the real numbers. This, on the relationist analysis, entails that there is an actual time existing in between t and $t′$. Again, the argument applies quite generally and relies on no contingent assumptions, so relationism entails that time is continuous as a matter of necessity.

Now it might appear that this particular proof can be blocked, as follows: a restriction introduced earlier was that if a possibility of change is to be sufficient for the actual existence of a time, then this possibility must be compatible with no disturbance in the actual temporal relations between events. Let us make the initial supposition that time is discrete. This provides us with a natural way of construing the metric of time: the interval between

---

[20] Newton-Smith makes a similar remark concerning spatial vacua: places cannot simply be described as *physical* possibilities of location because it may be physically necessary that a place is unoccupied. I find it puzzling that Newton-Smith doesn't consider the corresponding problem concerning time.

any two events is determined by the number of discrete instants between them. If there are no instants between two events, then those events are contiguous. Consider now two contiguous events. Is it possible for there to be an event between them? Yes, but only at the cost of disturbing their temporal relationship, for in any world in which there is an event between them they are *not* contiguous. So, given our restriction, the possibility of an event between two contiguous events does not, on relationist grounds, entail the actual existence of a time between them. So it would appear that we cannot, after all, construct a relationist *reductio* of the supposition that time is discrete. Consequently, any relationist 'proof' of time's continuity must be flawed.

However, the point was to show whether, *starting from no assumptions* about time's topology, we could generate any conclusions about topology from relationism alone, not necessarily to construct a relationist *reductio* of non-standard topologies (although such a *reductio*, if successful, would also establish the connection between relationism and topology). To start with the assumption that two events are contiguous is, of course, to start with an assumption about topology. So the result above still stands.

Linear versus Closed Time

Time is linear if, and only if, instants of time are isomorphic to the collection of points on a line, and closed if, and only if, they are isomorphic to the collection of points on a circle. So in closed time, every instant is both before and after every other (including itself). 'Before' and 'after' are asymmetric and irreflexive in linear time, symmetric and reflexive in closed time. This is the basic conception of closed time. In fact, however, the closed-time hypothesis is often presented as including the proposition that time is only finitely extended.[21] It is certainly the case that, if we had empirical grounds for the belief that time was *both* finite and unbounded, then we would also have reason to believe that time was closed. If, on the other hand, we had reason to suppose that time is infinitely extended, then any empirical observation we made would be compatible both with the hypothesis that time is linear and with the hypothesis that time is closed. There is a simple spatial analogy here: close inspection fails to reveal whether one has come across an infinitely extended line or an infinitely extended circle.

Now although relationism does not directly entail that time is linear, it does entail that time is infinitely extended, and this rules out one version of the closed-time hypothesis. The proof here is very similar to the one we gave for

---

[21] This is how it is presented both by Newton-Smith and by S. W. Hawking (*A Brief History of Time* (New York: Bantam, 1988)).

the boundlessness of time. Let t be the time at which event e occurs. It is always possible that there should exist an infinitely extended series of distinct events, of which e is the first (last) member. It follows that, on the relationist analysis, for any t, there will always be a time infinitely later (earlier) than t. So time is infinitely extended as a matter of necessity. Finite closed-time worlds, the kind often envisaged in discussions of closed time, are therefore ruled out. But, if we live in a world of infinite temporal extension, no empirical considerations could count as weighing in favour of infinite closed time as against infinite linear time, whereas we *could* imagine empirical considerations weighing in favour of *finite* closed time as against infinite linear time, or vice versa. We might discover, for example, that physical laws only allowed a finite degree of diversity—i.e. allowed only a finite number of qualitatively distinct states of the universe. If we held to the principle of the identity of indiscernibles, we would also regard this as establishing that time itself was only finitely extended. If we made the further step of accepting a physical hypothesis which contained a 'no boundary condition',[22] then we would have grounds for accepting the hypothesis of closed time. But the first step of this is ruled out by relationism. So relationism has the following rather interesting consequence for the linear versus closed-time dispute: it is not an empirical dispute. Given the infinite extension of time, a decision between closed and linear time would have to be based on a priori considerations.

Non-branching versus Branching Time

Time has a non-branching future if, and only if, the following principle is true:

$$(F_n(p) \,\&\, F_n(q)) \to F_n(p \,\&\, q).^{23}$$

That is, if p will be the case in n units' time *and* q will be the case in n units' time, then it will be the case in n units' time that p *and* q. A non-branching past can be defined in similar fashion, namely, if, and only if, the following principle is true:

$$(P_n(p) \,\&\, P_n(q)) \to P_n(p \,\&\, q).$$

A branching time-series is one in which one or both of the above principles is false. A diagram will help to make this clear (see Figure 1). Figure 1

---

[22] See Hawking, *A Brief History of Time,* 136–7, 145 f.

[23] A tense logic for non-branching time which does not involve metric tense operators is given in A. Prior, *Papers on Time and Tense* (Oxford: Clarendon Press, 1968). I use such operators here for simplicity.

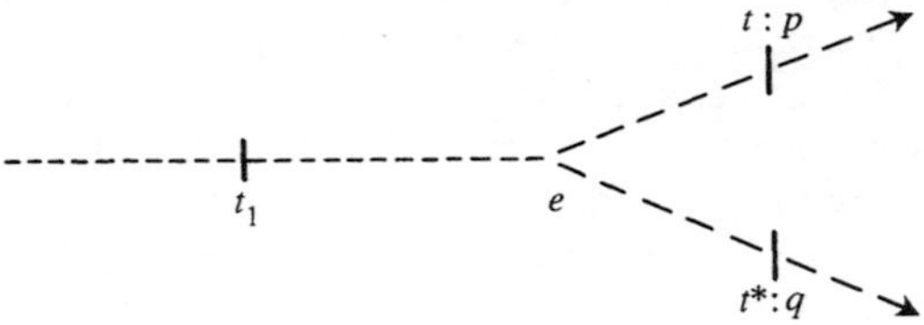

Fig. 1

represents a time-series which branches at one point—after the occurrence of some event e—to form two disjoint time-series. At t_1, both p and q lie in the future. But at t, although it is the case that p (now), it is *not* the case that q (now), since q occurs at t^*, which is temporally unrelated to t. So the first principle above fails.

Now the relationist analysis of instants, as formulated in (B′), entails that time is non-branching. The analysis, recall, went as follows: a time t, n units before/after an actual event e, is just the collection of actual and possible world-occurrences of actual or possible events n units before/after e. Suppose now that time branches after e, as above. t and t^* are times in different time-series which both occur n units after e. But if a time n units after a certain event is just the collection of actual and possible world-occurrences of events n units after that event, then t and t^* are identical, for they correspond to the same collection of actual and possible world-occurrences. So t and t^* cannot, after all, be in disjoint time-series. Time, therefore, cannot as a matter of necessity have a branching future. And, by a similar argument, it cannot have a branching past either.

What this shows, in fact, is that the choice of e for the relationist analysis cannot be an arbitrary one. It could only be arbitrary if time was of necessity non-branching. A time must be defined, not just by its relationship to a given actual event, but by its relationship to *all* actual events. The relationist can then accommodate the possibility of branching time by amending his reductionist analysis as follows: let $e \ldots e_n$ be the collection of all actual events, and let $R_1 \ldots R_n$ be the collection of relations which t bears to the members of $e_1 \ldots e_n$. Then t is just the collection of actual and possible world-occurrences of actual and possible events which bear $R_1 \ldots R_n$ to $e_1 \ldots e_n$. (Call this amended analysis (B″).) Now since t and t^* above will not bear $R_1 \ldots R_n$ to the same set of events (t will be temporally related to certain events which t^* is temporally unrelated to, and vice versa) they will not correspond to the same collection of world-occurrences. Relationism, it seems, does not have any consequences for whether time is branching or not.

4. CONCLUSION

To summarize the results of the previous section, relationism—as characterized by (A) and (B'')—entails the following:

(1) time is of necessity unbounded—i.e. has no beginning or end;
(2) time is of necessity continuous;
(3) time is of necessity infinitely extended;
(4) the debate between the hypothesis of closed time and that of linear time is not (to any degree) an empirical one.

Relationism thus entails that what we called the traditional view of time's topology is largely correct. The support is not total, because relationism does not completely rule out closed or (when suitably amended) branching time.

This is a surprising result. How should we respond to it? The simplest response is to insist that the modality in (A) and (B'') should, after all, be construed as a physical one. This would certainly break the connection between relationism and topology, but it also conflicts with the intuitions which formed part of the motivation for relationism. One was the intuition that temporal vacua—even those which are *determined*, and so in the circumstances physically necessary—are possible. The second was that times can be the subjects of modal statements which invoke physically impossible situations.

A second response is to limit the possible worlds, not to physically possible worlds, but to those worlds which share their temporal topology with the actual world. That is, there exists a time n units from e, if, and only if, it is *compatible with the actual topology of the time-series* that there should exist an event n units from e. Suppose that time actually has a bounded future and past. Then it is not logically possible, compatibly with this condition, that there should exist an event lying infinitely distant from any actual event. Now one problem with this response, which deals with our anomaly by brute force, is that it seems to be quite arbitrary. What, on the face of it, has a condition concerning topology to do with the basic proposition that times are possibilities of change? A second difficulty is that the response entails an essentialist thesis concerning the actual time-series, namely, that the actual time-series could not possibly have been part of a time-series which exhibited a quite different topology. So, if time is actually bounded in both directions, then it could not have been part of a time-series which was infinitely extended. It is still allowed that we can coherently talk of a possible world in which time has a different topology. But that world and this world have no times in common. This, I submit, is counter-intuitive.

That leaves us with two options. The first is to side with the traditionalists and concede that (at least some) topological theories of time are philosophical in character—or rather, that they could not be empirical if relationism is true, and whether it is true or not is purely a matter for philosophical debate. But then we would be left with the rather anomalous result that some topological properties are determined and others not. The second is to accuse relationism of poaching on empirical preserves by anticipating the results of empirical investigations. And even if the topology of time does after all turn out to be the standard one, this would still only be a contingent fact about the universe. Since relationism entails that time is *of necessity* unbounded, continuous, and infinite, defenders of the current orthodoxy will reject relationism.

The second response is the one I favour, but I recognize that defenders of a venerable tradition may welcome the results of this paper as directing us back to the Aristotelianism from which we should never have strayed.

POSTSCRIPT

Since writing 'Relationism and Temporal Topology' I have come to feel increasingly distanced from a number of theses advanced in it. I would like to record my changes of mind here, together with some further reflections on the topic of the essay.

To ask whether the topology of time should be a matter for the scientist or the philosopher is to invite a picture of their activities as clearly distinct. But only a parody portrays the scientist as merely accumulating facts and making generalizations on the basis of them. The 'raw' data, and the formulae which putatively describe them, require interpretation. Abstract ideas and a priori principles are never far away. For example, if we find that the result obtained in measuring the speed of light is independent of the relative speed of the light source, then whether or not we take this to imply the relativity of simultaneity will depend on how willing we are to give up certain preconceptions. I see no reason to suppose either that the scientist will be unaffected by a priori principles or that the philosopher will be indifferent to empirical findings. So the crucial question with regard to the topology of time, then, is not whether it is physics or metaphysics, but whether there are any sound a priori arguments for time's having a certain topological structure.

Now, as I indicated in the paper, I had already formed an opinion on this question: that the answer was 'no'. I then went on to argue that relationism (as I construed it) entailed certain topological properties, and that this meant that relationism was improperly anticipating empirical findings. I now think that this was an unfair criticism. After all, what should make anyone think that the topology of time cannot be a purely a priori matter if not the absence of sound a priori arguments for a certain topology? *If* there is a valid argument from relationism to the standard topology, as I suggested there was, then we can hardly complain if relationists adopt that topology. We cannot regard relationism as suspect on *those* grounds.

However, and this is the main point of this postscript, I now think that the proofs I offered were suspect. But in showing how they are suspect, we obtain some interesting results.

Recall the relationist formula:

(A) $(\exists t)(R_n(t, e)) \leftrightarrow \Diamond \exists x(\text{Event}(x) \& R_n(x, e))$.

I characterized (A) as an attempt to state the necessary and sufficient conditions for the existence of a time, but in fact all (A) provides is a schema. We can only advance to a statement of conditions when we substitute a particular relation, or set of relations, for 'R', define the properties of this relation, and state which series of numbers n is to range over. It would be natural to replace 'R' by 'Before' ('Before$_0$' can be used instead of 'simultaneous with'), but these are only suitable if we make an assumption about time's dimensionality—a topological property I didn't discuss in the paper. If time is one-dimensional, then we can describe all the temporal relations between times and events using just the binary relation above. If, however, time is *two-dimensional* (an idea which, as Murray MacBeath has shown in his contribution to this volume (Essay XI), can be given empirical content) then we shall need to employ additional relational terms, in order to locate the events in the second dimension. The relationist will need to decide, then, which view of the dimensionality of time is going to be built into his analysis: one-dimensional, two-dimensional, or n-dimensional. Suppose the relationist just opts for the 'Before' relation, then he commits himself to one-dimensional time. This choice is not forced on him—he could opt for two dimensions—but a decision has to be made.

The relationist must also decide what properties are to be ascribed to the relation(s) he chooses. For example, is 'Before' to be regarded as globally asymmetric? The reason this and similar questions have to be settled is this. Suppose that, in some possible world, time is closed. That is, any event in that world is both before *and* after any other event in that world: 'Before', in closed time, is globally symmetric. So for any event e in this, the actual, world, it is logically possible for there to exist another event, e^*, such that e^* is n units before e and n units after e. But that, on the relationist picture (as I represented it), entails that there is an actual time, t, such that t is n units before e and n units after e. For such a relationship to exist, however, time must actually be closed. So relationism appears to license the inference from the logical possibility of closed time to the proposition that time is *actually* closed. This strange result can be avoided if the relationist is prepared to stipulate what properties are to be ascribed to the 'Before' relation in his analysis. If it is defined as being globally asymmetric, then the relationist has decided in favour of linear time, although he has not ruled out the possibility of time being closed.

Consider now the index n. Unlike 'R' this is not a name but (in any complete statement of (A) in which all the quantifiers are made explicit) a bound variable, ranging over numbers. In the proof of time's continuity I assumed that n ranged over the real numbers. The relationist, however, need not accept this. Suppose time is in fact discrete. Then it would be appropriate to restrict the range of n to the integers. If time is dense, then n should range only over the rationals. The relationist needs to settle the issue of which number-series defines the range of n. But to do this he must settle on a particular topology: discrete, dense, or continuous. Continuing this theme: should the number-series be infinitely or only finitely extended? This, again, is determined by a decision over whether time is finite or infinite (but see below).

This leaves us with the issues of unified versus disunified and branching versus non-branching time. On the issue of unity, at least, the relationist is apparently faced with a problem. Note that (A) makes singular reference to a particular event. If singular reference to an entity entails causal contact between that entity and the use of the referential term, then the relationist will have difficulties extending (A) to cover disunified time. Where time is disunified, there are two or more time-series which are temporally unrelated to each other. Since causal contact entails temporal relations, the occupant of any one time-series will be causally isolated from the occupant of any other time-series. On any causal theory of reference, one cannot therefore make singular reference to an event in a time-series disjoint from our own. The problem, however, seems to me to be a superficial one. All the relationist need show, in the statement of the necessary and sufficient conditions for the existence of a time, is how the existence of a time is related to possibilities of temporal location. The purpose of (A) is not the construction of a *particular* time, but rather the representation of the *form* of such a construction. Terms like 'e' should therefore be replaced by bound variables.

To summarize the main points of the discussion: the relationist cannot regard (A) as a statement of the necessary and sufficient conditions for the existence of a time until he specifies: (1) the relation(s) represented by 'R'; (2) the logical properties of that relation; (3) the number-series over which n ranges. In making these specifications the relationist is deciding whether time is: (*a*) one-dimensional, two-dimensional, etc.; (*b*) linear or closed; (*c*) discrete, dense, or continuous; (*d*) finitely or infinitely extended; (*e*) bounded or unbounded (this is a consequence of other decisions; for example, if time is closed, then it is unbounded). No particular topology is forced upon the relationist, however: he can make his decision on the basis of empirical considerations. In addition, his decision carries no implications for the possibility or otherwise of non-actual topologies. For the actual time-series, he may provide a version of (A) which reflects time's continuity, but for another time-series, he may provide a version on which time comes out discrete.

In the paper, I attempted to show that '*starting from no assumptions* about time's topology, we could generate . . . conclusions about topology from relationism alone'. I now want to reverse this thesis and assert instead that relationism is an incomplete theory *unless* it makes explicit assumptions about the topology of time. What one might call the 'cautious' relationist will offer a number of analyses suitable for various topologies, just as the 'cautious' tense logician produces different tense-logical systems without endorsing any one system as the correct one. But we may have a relationist who throws caution to the winds and refuses to put constraints on, for example, the series of numbers ranged over by *n*. Call this theorist the 'liberal' relationist. The liberal relationist takes the view that time is nothing more—*or less than*—the aggregate of possibilities of change. Such a theorist would regard time as something whose structure could partly be revealed through a priori reflection. To capture all the possibilities, the liberal relationist will insist that *n* must be allowed to range over the infinite series of real numbers. On this view, time will of logical necessity be continuous and infinitely extended. What is interesting about the liberal relationist position is that it introduces an asymmetry between the questions of extension and microstructure on the one hand, and questions of linearity, dimensionality, branchingness, and unity on the other. The liberal attitude determines the decision over *n* but not, as far as I can see, any decision over the identity and properties of R. Liberal relationism is a curious doctrine, but not, as far as I can see, an incoherent one.

The cautious relationist, though freed of any commitment to the traditional view of time's topology, is still committed to the essentialist view outlined at the end of my paper: that the topological structure of a particular time-series is an essential property of that series; it could not, in some other possible world, have had a different topology. For example, if the actual time-series has, as ordered by the 'Before' relation, a linear topology, that very series could not have been closed. I described this consequence as counter-intuitive, but it is one which the relationist might be prepared to accept. It becomes a far less acceptable consequence, however, if he subscribes to a certain actualist view of times, namely, that the only times which exist in this or any possible world are *actual* times: we cannot coherently talk of merely possible times. Consequently, any two possible worlds will have a time-series in common. Combining this actualist position with the essentialist one leads to the conclusion (characteristic of the traditional view) that time has the topological structure it does as a matter of logical necessity. This would be a striking result, for then we would have the possibility of the relationist deciding on empirical grounds in favour of a particular topology for time and yet holding that no other topology would have been logically possible. Truths about, for example, the continuity of time would then be an interesting species of necessary a posteriori truth.

This is not the place for a full discussion of actualism over times, but I would like to indicate briefly what motivation there could be for such a view, and to show that the relationist has in fact good reason for adopting it.

To deny actualism over times is to take the view that it is coherent to suppose that two possible worlds could fail to have a time-series in common. Under what circumstances would two worlds fail to share a time-series? For the relationist, it would be natural to provide an answer in terms of the events which occurred at those worlds. The relationist will allow times to have different contents at different worlds, but he must, I think, insist that a token event is essentially connected to a token time-series (though not necessarily to a token time). That is, it is not possible for a token event to occupy a different time-series from the one it occupies in this world. This is not such an implausible principle. When we say that the Fire of London could have happened at a different time, we mean, surely, that it could have occurred *earlier* or *later* than the time it actually did. We do not mean that that token event could have occurred at a time unrelated to the token time actually referred to by '1666'. The relationist is committed to the principle that token events are essentially tied to a token time-series because for him times are identified in terms of their relations to token events. Is it also true that, for the relationist, *any* token individual (where this could mean event, material object, subatomic particle) is tied to a token time-series? Perhaps not, but this extension of the principle is a plausible one, and it would be strange to treat events differently from other particulars in this respect. Let us assume that the relationist accepts this extension. Then having no token individuals in common is a necessary condition of not sharing a time-series.

Can we allow two possible worlds to have no individuals in common? It depends upon which theory of modality we accept. If we reject modal realism, then we have to locate the facts which make modal statements true either among such abstract entities as propositions or among the concrete elements of the natural world. The latter course, is, I take it, the more attractive option (since, arguably, propositions are themselves only definable in modal terms[24]). How can we construct possibilia from concrete elements? Combinatorialism provides a simple and plausible answer: a possible state of affairs is simply a recombination of actual individuals and actual properties or actual relations.[25] For instance, suppose it is actually the case that Fa and Gb, where 'a' and 'b' are names for individuals, and 'F' and 'G' are names for properties. Then we can construct purely possible states of affairs simply by recombining the elements: Fb and Ga. Now if we are combinatorialists, we can allow purely possible states of affairs, *but we cannot allow purely possible individuals*. That is because possibilia can only be constructed out of *actual* individuals. So the conjunction of combinatorialism (or any modal theory which entailed actualism over individuals) and relationism entails actualism over times. Relationism does not, by itself, entail such actualism, but the relationist, nevertheless, may have good reason to hold it.

[24] See e.g. Lewis, *On the Plurality of Worlds*, 53–5.

[25] See e.g. William Lycan, 'The Trouble with Possible Worlds', in M. J. Loux (ed.), *The Possible and the Actual* (Ithaca, NY: Cornell University Press, 1979), 274–316; D. M. Armstrong, *A Combinatorial Theory of Possibility* (Cambridge: Cambridge University Press, 1989).

X

THE BEGINNING OF TIME

W. H. NEWTON-SMITH

Infinity lies in the nature of time, it isn't the extension it happens to have.

(Wittgenstein, *Philosophical Remarks*)

1. BEGINNINGS AND ENDINGS

Could time have had a beginning? At first glance and perhaps even at second glance posing this question seems to set us on the well-travelled road to antinomy. For instance, if we suppose that time had a beginning, our normal linguistic habits lead us, seemingly inexorably, to talk inconsistently of time before that beginning. To suppose, on the other hand, that time could not have had a beginning will lead us to conclusions which while consistent are unpalatable. For given that change might have had a beginning we are committed to thinking of any world with a first event as a world with endless aeons of empty time. And our reluctance to embrace empty time leads us to be chary of admitting this emptiness. For as we will see in Section 4 of this essay, the posit of time before a first event cannot play an explanatory role.

There are substantial issues at stake here. Before we can focus on them we need to introduce some routine clarification of the notion of a beginning of time. The notions of beginning and ending are most frequently applied to things in time, and it might seem that in applying these notions to time itself we are in danger of entering a morass of nonsense. Indeed it has seemed to some writers that to pose the question of whether or not time had a beginning is to commit a category error. On this view it is only of things in time that one can legitimately ask of their beginnings and endings. For example, that it begins to snow entails that at one time there is no snow falling and that at a later time there is snow falling. On this construal, to say that time began involves making either the contradictory assertion that, at one time, time did not exist, and then at a later time, it did exist; or the absurd supposition that

W. H. Newton-Smith, *The Structure of Time* (London: Routledge & Kegan Paul, 1980), ch. 5
Reprinted by permission of Routledge.

some super-time exists relative to which time itself can be said to begin. But in point of fact the only moral to be drawn here is the entirely insubstantial one that we cannot think of the beginning of time on an exact parallel with the beginnings of things in time. What we should have in mind when we say that time began is something akin to what we have in mind when, for instance, we say that the natural number series begins with zero. This means that there is no natural number coming before zero when the numbers are taken in the standard order. It is on this model that I will understand the notion of beginning time. That is, the claim that time began is to be understood as the claim that the set of temporal items has a first member under the basic temporal-ordering relation. If we take the set of temporal items as the set of instants and the basic relation as that of being temporally before, the hypothesis that time had a beginning is the hypothesis that there is an instant such that no other instant is before it. And the hypothesis that time had no beginning is the hypothesis that for every instant there is a distinct earlier instant.

2. ARISTOTLE, SWINBURNE, AND TENSES

The thesis that time could not have had a beginning has been defended by appeal to what are alleged to be logical truths involving tenses. In his recent book *Space and Time*, Richard Swinburne has argued as follows:

Time . . . is of logical necessity unbounded. Before every period of time which has at some instant a beginning, there must be another period of time, and so before any instant another instant. For either there were swans somewhere prior to a period *T* or there were not. In either case there must have been a period prior to *T* during which there were or were not swans.[1]

In what follows I will make use of the notion of a tense-logical truth.[2] By a tense-logical truth of such and such form I mean that any proposition of that form would be true at any time in any possible world. I will also make use of some simple tense-logical notation, write '*P*—' for the past-tense operator 'it was the case that', and '*F*—' for the future-tense operator 'it will be the case that'. I shall take it that in our semantics, simple tensed

[1] Swinburne, *Space and Time* (London: Macmillan, 1968), 207 Actually Swinburne offers the future analogue of this argument and claims the past analogue of his argument holds. In the quoted passage, I have transposed his argument to the past-time case Swinburne claims the argument given above establishes unboundedness. To establish that time is non-ending and non-beginning requires the additional premiss that time is not closed. Swinburne argues that time of logical necessity is not closed. In what follows I assume we are dealing with non-closed time and hence treat his argument as an attempt to show that time of logical necessity is non-beginning.

[2] [This notion is introduced on p. 63 of Newton-Smith's *The Structure of Time*, from which this essay is drawn.—*Editors*]

propositions are assigned the value T (true) or F (false) at each time and that the complex proposition Ps is assigned T at some time if and only if there is some earlier time at which s is assigned T. Fs is assigned T at some time if and only if there is a later time at which s is assigned T. In these terms we can clearly distinguish between the propositional forms $P\sim s$ and $\sim Ps$, the former being the past-tense assertion of the denial of s, and the latter being the denial of the past-tense assertion of s. For $P\sim s$ can be true at time t only if there were prior times at which s is false. But $\sim Ps$ can be true at time t if either there are prior times and at all those times s is false; or there simply were not times prior to time t.

Consider now Swinburne's premiss: There were swans or there were no swans. The problem here is that the English sentence 'There were no swans' is Janus-faced. It can be used either to deny the past-tense proposition 'There were swans', which we can write as '$\sim Ps$'; or to assert the past-tense version of the denial of 'There are swans', which we can write as '$P\sim s$'. Taking this disjunct of the premiss in the latter way renders the argument question-begging; taking it in the former way does not give us a valid argument at all. Suppose we construe the premisses as '$Ps \vee P\sim s$'. Granting the truth of this, at some time t, allows us to argue from either disjunct to the existence of time before t. But why should we accept this premiss as having the status of being tense-logically true? The reductionist will argue that this cannot be a tense-logical truth on the grounds that, if there is a first event, the proposition is false at the start of that event. Thus, unless some independent argument is advanced for thinking that time of logical necessity had no beginning, we have no grounds for regarding this as being a tense-logical truth and hence the argument is question-begging.

That Swinburne offers no reason for accepting the crucial premiss suggests he is taking it in the other formulation: namely, as '$Ps \vee \sim Ps$'. In this case, as we have a substitution instance of the law of the excluded middle we (or at least non-intuitionists) might be expected to accept it as a tense-logical truth. However, we cannot now argue from the disjunct '$\sim Ps$' to the existence of prior times. For on the understanding of the tense operators that Swinburne needs for his argument '$\sim Ps$' *might be true in virtue of the absence of prior times*. It is only by adding as an additional premiss: '$\sim Ps \rightarrow P\sim s$' that the argument goes through. But this additional premiss is logically equivalent to our other formulation and for that reason renders the argument again ques-tion-begging.

Thus, on neither interpretation does Swinburne's argument go through. The *appearance* of an argument is generated by the fact that in normal contexts we do allow an inference from $\sim Ps$ to $P\sim s$, *for the obvious reason that we normally operate under the implicit assumption that there were prior times*. Under that assumption, the inference is innocuous. If we wish to represent this standard inference pattern in such a way as to avoid an implicit commitment to non-beginning time we can use, say: $\sim Ps \rightarrow (P(s \rightarrow s) \rightarrow P \sim s)$. By putting any tautology in the second antecedent we block the inference from $\sim Ps$ to $P \sim s$ at t except under the assumption that *something* was true; i.e. that there was time before t.

Swinburne's argument is rather reminiscent of the following argument of Aristotle's for unbounded past and future time:

> Now since time cannot exist and is unthinkable apart from the moment, and the moment is a kind of middle-point, uniting as it does in itself both a beginning and an end, a beginning of future time and an end of past time, it follows that there must always be time.[3]

This amounts to the claim that it is a necessary condition of a moment's ever being present that there are at that time past and future moments. This claim can be represented by the following pair of tense-logical postulates:

(1) $q \rightarrow FPq$;
(2) $q \rightarrow PFq$.

These postulates express the thought that for any present-tense proposition which is true at a time there is then a past time during which the proposition was going to be true and a future time at which it will have been true. Together these postulates, which have commended themselves to others,[4] entail Aristotle's thesis that there is no present that is not, so to speak, flanked by a past and a future. It follows from these postulates that time is non-ending and non-beginning. However, we cannot establish that time is thereby of logical necessity unbounded without first establishing that these postulates are tense-logical truths. But establishing that requires establishing just what is at issue.

One certainly does feel that there is some non-contingent link between a proposition's now being true (or an event's now being present) and its going to be that it was true (or a present event's being past in the future) and its having been going to be true (or a present event's having been future in the past). This feeling gives initial plausibility to Aristotle's argument. One way we can do justice to this feeling without thereby committing ourselves to necessarily unbounded time is to adopt the following technique and take as tense-logical truths the following pair of weaker postulates:

---

[3] Aristotle, *Physics*, vIII. 251^b 19–23.
[4] In this regard see J. N. Findlay ('Time: A Treatment of Some Puzzles', *Australasian Journal of Philosophy*, 19 (1941): 216–35), who opts for a stronger postulate than (1) that all events, past, present, and future, will be past He would presumably also opt for the postulate that all events (past, present, and future) have been future, which is stronger than (2).

(1) $q \rightarrow (P(s \rightarrow s) \rightarrow PFq)$;
(2) $q \rightarrow (F(s \rightarrow s) \rightarrow FPq)$.

Given these postulates, one can pass from p's being true to its going to be that p was true only in conjunction with the assumption that there is a future. This additional condition is captured by applying the future-tense operator to any tautology. For if there is a future, $F(s \rightarrow s)$ is bound to be true, and if there is no future, it is bound to be false. In Aristotle's idiom we have advanced the weaker claim that nothing excepting a first or a last moment of time can be present unless it is a mean between the past and the future.

One can characterize the style of argument we have been considering as follows. We are offered a sentence schema involving tense operators—a schema that *seems* to reflect an inference pattern we standardly employ. The sentence schema is taken to be what I called a tense-logical truth and it is concluded that time has to have whatever topological properties the schema's being a tense-logical truth would entail. These arguments seem plausible as one may think that grasping the inference patterns represented by the schema is part of what it is to grasp the sense of the tense operators. Swinburne, for instance, speaks of the ordinary concept of time. One imagines that he has in mind the thought that our ordinary concept of time involves concepts of the past, present, and future the content of which is in part represented by the inference patterns in question. However, we can represent the rules governing the usual tense operators in ways that will do justice to our usual inference habits without implicitly attributing a particular topological structure to time. As we saw, one might equally offer the following schema: $p \rightarrow (F(p \rightarrow p) \rightarrow FPp), p \rightarrow (P(p \rightarrow p) \rightarrow PFp)$. The general moral that is to be drawn is the following. We cannot establish that time of logical necessity has a given topological property by appeal to allegedly tense-logical truths. For any such argument will need to be supported by a proof that such allegedly tense-logical truths are indeed tense-logical truths and this in turn will require a proof that time of logical necessity has the topological property in question.

3. KANT, INDEFINITE EXTRAPOLATION, AND POSSIBILITY

In the Antinomies Kant purports both to derive a contradiction from the assumption that the universe had a beginning in time and to derive a contradiction from the assumption that the universe had no beginning in time. It is not always noted that in both cases the argument takes place under the assumption that time had no beginning. Kant appears to take it that the following consideration adduced in the Aesthetic establishes the non-beginning of time:

> The infinitude of time signifies nothing more than that every determinate magnitude of time is possible only through the limitations of one single time that underlies it. The original representation, *time*, must therefore be given as unlimited.[5]

If I understand this argument, and I am not at all sure I do, it may come to this. Any particular duration (determinate magnitude of time) is the duration containing such and such instants and not containing all other instants (the limitations of one single time). If we add the implicit premiss that for any duration there is a larger possible duration, Kant seems to be claiming that time must be thought of as infinite in order to accommodate the unending sequence of longer and longer durations.

Given a particular metrication of time, if for every event or process which has occurred, some event or process of greater duration had occurred, we should be committed to thinking of time as having infinite past duration, *under that metrication*. However, if we came to think that there was a first event and if our metrication assigned the interval of time elapsed since that event some finite value, we would not be committed thereby to thinking of time itself as pre-existing, so to speak, this first event. We can assign a well-defined finite value to the duration from the beginning of the first event until now without presupposing time prior to that event. To put the point colourfully—time must be thought of as at least as big as any process or sequence of processes in time. But we need not think of it as actually bigger than any actual sequence of processes that occur.

The particularly unconvincing character of the arguments of Swinburne, Aristotle, and Kant tempts me to offer the following *ad hominem* explanation of what brings many to think that there must be some cogent argument for non-beginning time. Our uncritical thinking about time is influenced by the technical devices we use for representing the temporal aspects of things. We often label moments at which events occur by elements of some number system. Some auspicious event like the birth of a prophet is assigned, say, zero. Earlier and earlier events are assigned larger and larger negative numbers; later and later events are assigned larger and larger positive numbers. An event taken to be the first is assigned some particular large negative number. As the number sequence itself used can be further extrapolated we are seduced into thinking that there must be times corresponding to these numbers. However, the extrapolation within the representing system does not guarantee that those extrapolated elements of the system have referents. To

---

[5] I. Kant, *Critique of Pure Reason*, trans. N. Kemp Smith (London: Macmillan, 1929), 75 (432=B47).

use an analogy—suppose an interest in a scientifically useful concept of temperature leads us to define temperature in terms of molecular motion and to set − 273.16°C as the temperature of a body whose constituent molecules are not in motion. The mathematical system of the real numbers used to represent the temperature of bodies admits extrapolation below − 273.16°C. But these lower numbers cannot be thought of as designating lower temperatures which as a matter of fact no body possesses. For, by definition, there is no temperature lower than − 273.16°C. I am not here suggesting there is an exact analogy with time. For it is not clear that talk of time before a first event is simply incoherent in the way that talk of lower temperatures than that of a body with no molecular motion would be.

Quinton has recently argued for the infinite extent of time by claiming:

The infinite extent of (Euclidean) space cannot be invoked to support the infinite size of the material world that occupies it in the way that its infinite divisibility was used to establish the infinitely divisible. At however remote a distance we have found matter to be present it is always possible to describe matter further off still. But it does not follow that there is any matter there. As long as there are some material things we can set up a system of spatial characterisations which allows for the description of possible things at any distance whatever. But it is a contingent question whether these positions are occupied, whether these possibilities are realised. Similarly with time. It has no beginning in the sense that there is no date an earlier date than which cannot be significantly described. But there may be a date at no time earlier than which anything was happening at all. Infinity, we might say, is a necessary feature only of systems of description, not merely those which contain numbers but any which contain such transitive asymmetrical relations as 'smaller than' and 'further than' and 'earlier than'. But this entails neither that there is no limit to the minuteness or remoteness of actual things nor, *a fortiori*, that there are infinitely small or infinitely remote things.[6]

Nothing very significant would be established if it could be shown that we are committed to the non-beginning of time by our 'system of descriptions', or by 'the grammar of our determinations of time'.[7] For it only prompts the more substantive question as to whether we should continue to employ that particular system of descriptions or grammar. We may have developed our system of descriptions without taking account of certain possibilities, i.e. that there was or might have been a first event, and the viability of that system of descriptions may depend on that possibility not being realized. In any event Quinton has not established that this is a feature of our system of descriptions. We cannot argue for the non-beginning of time on the grounds that we use a transitive asymmetrical relation in giving temporal order. For transitive asymmetrical relations can be defined on finite sets.

---

[6] A. Quinton, *The Nature of Things* (London: Routledge & Kegan Paul, 1973), 88.
[7] L. Wittgenstein, *Philosophical Remarks*, trans. R. Hargreaves and R. White (Oxford: Blackwell, 1975), 309.

It is not infrequently supposed[8] that one can argue from the conceptual connection between time and the possibility of change[9] to the non-beginning of time along the following lines. Suppose that as a matter of fact there was a first event, E_0, which occurred at time t_0. While E_0 was the first event, it might not have been. It was true that at time t_0 there might have been events occurring at times before t_0. Hence, it is argued, there were times before t_0 at which things might have occurred, but did not do so. The crucial premiss is the claim that there might have been events before t_0. If this is taken to mean that there were times before t_0 at which things might have happened, the argument is question-begging. For what is at issue is whether there were times before t_0. However, the claim is more plausible if construed as follows. In some possible world different from this world, E_0 is preceded by other events. In that possible world there are times before t_0. Hence it is true in the actual world that there *might have been* times before t_0. On this construal, we need some additional premiss to license the inference from the possibility of times before E_0 in the actual world to the existence in the actual world of times before E_0.

The above argument turns on the assumption that the possibility of events before a given event is sufficient to establish the actuality of times before the time of the given event. However, neither the logical nor the physical possibility of the occurrence of an event is a sufficient condition for the actual existence of a time at which the event might have occurred. Hence the assumption on which the argument in question turns is not tenable.

In a somewhat similar vein Peirce argued for the non-beginning of time:

If on a Monday an idea be possible, in the sense of involving no contradiction within itself regardless of all mere circumstances, then it will be possible on Tuesdays, on Wednesdays, and on Fridays; in short it will be possible forever and ever, unless the ideas of the circumstance should come into definite rational contradiction to the idea in question. Consequently, mathematical Time cannot have an arbitrary beginning nor end. For it is a possibleness; and what is possible at all is possible without limit, unless there be some kind of a limit which comes into definite rational contradiction with the idea of Time.[10]

Peirce seems to be arguing that if there was a first instant of time, t, there would be a limitation on logical possibility. For logical possibilities would not obtain before time t. That is so. There was no time before time t and

[8] See in this respect Clarke's Fourth Reply in *The Leibniz–Clarke Correspondence*, ed. H. G. Alexander (Manchester: Manchester University Press, 1956), and Wittgenstein, *Philosophical Remarks*, 163, where he claims that 'empty infinite time is only the possibility of facts which alone are the realities'.

[9] [This connection is argued for in chapter 2 of *The Structure of Time.—Editors*]

[10] C. S. Peirce, *Collected Papers of C S Peirce*, ed. C. Hartshorne and P. Weiss (Cambridge, Mass.: Harvard University Press, 1935), 224–5.

hence no possibilities could obtain before time t. But this result is quite compatible with the principle, appealed to by Peirce, that what is logically possible at one time is logically possible at any other time. For we can hold that at time t it *was* possible that p, if we construe this not as there having been a time at which p might have been true but as p's being the case in some possible world in which there are times prior to t.

4. EMPTY TIME, LEIBNIZ, AND EXPLANATION

We have seen that the case for the necessary non-beginning of time is unconvincing. In general those who have agreed have argued by appeal to what I call Aristotle's Principle (or AP)[11] that time had a beginning if and only if change had a beginning. Aristotle's Principle cannot be defended in any strong form that would entail that talk of time before a first event was incoherent or vacuous. However, we can use an argument of Leibniz to support the thesis that it would be inadvisable to posit the existence of time before a first event.[12] For given that the actual world had a first event, one who posits the existence of time before that event has to acknowledge that there is a possible world exactly like this one except for its location in time, and is hence faced with the question: why did the world begin at the time it did begin? One who equates the beginning of time with the beginning of change is not faced with that question. Of course both are faced with the question why the world began with the event it did begin with.

It seems prima facie that there could be no answer to this question. Consider the usual sort of case where we might ask why some event occurred when it did. Suppose I ask why the Canadian election took place last autumn and not last spring. We might in part explain the autumn election by citing some particular facts (initial conditions) about our political leaders' ability to perceive conditions conducive to their self-preservation, and some general facts (lawlike regularities), say, about the inclination of politicians to strive for self-preservation. In part, at least, we explain the election's relative position in the temporal sequence of events constituting the history of our world by citing other things besides the position of the election that would be different in a world in which the election had a different position. But on the assumption that there was time before the first event, there are other possible worlds exactly like this one (assuming it to be a first-event world) except for location in time, and so, *ex hypothesi*, there would be no possibility

---

[11] [Aristotle's Principle is the claim that there is no period of time without change; see chapter 2 of *The Structure of Time.—Editors*]

[12] *The Leibniz–Clarke Correspondence*, 27.

of explaining why this world has the location it does in time by citing other differences between the worlds.

If it were necessarily true that everything is in principle capable of explanation, we might have an argument to support the thesis that there could not be time before change. As this is not a necessary truth, we do not have such an argument. However, the argument has some force. For it would certainly count against the postulation of time before a first event that it involves postulating a state of affairs (the world's beginning at a certain point in time) which might have been different (it could have begun at another time), but for which no explanation could be given. We have, other things being equal, a preference for not postulating states of affairs for which no explanation can be given. And, perhaps more importantly, we do not see what would count in favour of the postulation of empty time prior to a first event. There are no facts about the world which would be explained by such a postulation. It could only be a sensible postulate with explanatory force in the context of a theory according to which a world without time before its first event would have some other difference besides this difference from a world with time before its first event.

5. ARISTOTLE'S PRINCIPLE, EMPTY TIME, AND THE BEGINNING OF THE UNIVERSE

If my earlier discussion[13] succeeded in establishing that there is no incoherence in positing periods of time without change, the reductionist cannot appeal *tout court* to Aristotle's Principle, AP, in arguing that time before the first event is not a logical possibility. So it would seem that we cannot rule out a priori the existence of time before a first event, nor, given the failure of the arguments against boundedness, can we assert a priori the existence of time before a first event. We are left then with the question of what would be reasonable grounds for asserting the existence of time prior to a first event. For in keeping with our modified version of AP, the positing of periods of time without change is sensible only if that posit is part of a theory which fares better than its rivals in giving an account of observable change. It remains possible to argue that no explanatory end would ever be served by the posit of time prior to a first event. An examination of the structure of arguments adduced by cosmologists in support of the contention of a first event seems to support that conclusion. To such an examination I now turn.

---

[13] [Not included in this volume, though the argument in question is a variant of Shoemaker's argument in Essay IV.—*Editors*]

I have treated the notion of the beginning of the universe as unproblematic. If there was some incoherence in the very notion of a first event, we should have an argument for the non-beginning of time as a matter of logical necessity *without* a violation of Aristotle's Principle. Leaving Kant aside for the moment, the most frequently encountered argument for this conclusion is perhaps the following. It is permissible to talk about the beginning of processes in the world, but to transfer this talk to the world itself is to commit some form of category error.[14] Certainly we must take care in transferring to the world itself notions paradigmatically applied to subsystems of the world. For example, the notion of entropy raises certain problems of this kind. For if we have defined entropy as a property of a closed system, it is not clear how we can extend this notion of entropy to the entire universe. The notion of a closed system is defined in such a way as to make it unclear what could be meant by considering the universe itself as a closed system. However, the only moral to be drawn here is that care must be taken. The hypothesis that the universe had a beginning is conceptually unproblematic if it is taken as the hypothesis that the set of all past events had a first member.

Kant purported to find a contradiction in the notion of a beginning of the universe:

For let us assume that it has a beginning. Since the beginning is an existence which is preceded by a time in which the thing is not, there must have been a preceding time in which the world was not, i.e. an empty time. Now no coming to be of a thing is possible in an empty time, because no part of such a time possesses, as compared with any other, a distinguishing condition of existence rather than of non-existence; and this applies whether the thing is supposed to arise of itself or through some other cause. In the world many series of things can, indeed, begin; but the world itself cannot have a beginning, and is therefore infinite in respect of past time.[15]

Kant assumes that time could not have had a beginning and is arguing that if there was a first event there could be no explanation as to why it occurred at the time it did occur. Even if it were granted that everything must (logically) be capable of an explanation, it does not follow that there is an incoherence in the notion of a beginning of the universe. For one could deny equally the assumption which Kant has failed to justify, that time is of necessity non-beginning. If we take time and the universe as beginning together, so to speak, there is just no question as to why it began at the time it did begin.

More interesting difficulties arise if we consider the epistemological problem of what would constitute evidence that our universe had a beginning.

---

[14] See in this regard R. Harré, 'Philosophical Aspects of Cosmology', *British Journal for the Philosophy of Science*, 13 (1962): 104–19.

[15] Kant, *The Critique of Pure Reason*, 397.

Sceptics about induction are apt to feel particularly sceptical with regard to claims about the state of the universe a very long time ago. And those who are not simply sceptical about induction find their courage falters in the face of the degree of extrapolation involved in cosmological arguments. But if we are not sceptics about induction in general, we have no reason to falter, though we ought to have only a modest confidence in our beliefs concerning the very distant past. Setting aside these general doubts as not having particular relevance to the questions of the beginning of time and the beginning of the universe, there are, none the less, certain peculiarities in the sort of argument standardly adopted in support of the hypothesis of a beginning of the universe that do give rise to interesting and perhaps intractable epistemological problems. Arguments to support the hypothesis of a beginning of the universe typically have the following general structure. We suppose that we have some evidence that the radius, R, of the universe at time t is given by the following expression:[16]

$$R(t) = (at + b). \qquad \text{(a, b, constants)}$$

Density, $\rho(t)$, is then given by:

$$\rho(t) = e \frac{(dR)^2}{(dt)} \cdot \frac{K_2}{R}.$$

In this case there is some value t_0 of t for which $R(t)$ is zero in a model with $b = 0$. Under the assumption that matter has not, since time t_0, been created *ex nihilo*, the universe would have been at time t_0 in a state of infinite density (call this state S_0). We suppose further that we have evidence supporting other laws which are incompatible with the existence of a state S_0 of infinite density. Thus, it is reasoned, S_0 could not have obtained, and the first state of the universe must have been some state, S_+, occurring at some time t_+ at which $R(t_+)$ is small but not so small as to give the universe an impossibly high density.

I am not concerned with the question of evidence for the alleged lawlike regularities embodied in the expansion function; nor for the alleged lawlike regularities which are incompatible with the existence of infinitely dense states of the universe. Rather we are concerned with the identification of the state S_+ which we assume to have existed as the first state of the universe. Not infrequently we have evidence for a pair of laws which are discovered to have incompatible consequences in some contexts that fall within their scope. It is sometimes fruitful to seek a modification of one or other of the

---

[16] This universe characterizes the Einstein–de Sitter model of the universe. See J. North, *The Measure of the Universe* (Oxford: Oxford University Press, 1965), ch. 6, for an account of this and related cosmological models.

laws (given that they are highly confirmed over some range of data and do not have incompatible consequences over that range of data) which brings them into harmony. In the present case we are invited to correct the expansion function on the basis of the density laws. But this necessitates abandoning another well-entrenched belief, namely, the belief that at least at the macroscopic level all states of the universe are causal upshots of temporally antecedent states. In correcting the expansion law, we are forced to admit the existence of an uncaused state, S_+. One might argue the proper conclusion to be that S_+ is a state beyond which we cannot extrapolate on the basis of our present theories. It is more reasonable, the argument would run, to stick by our belief in the causal principle (applied macroscopically) than to stick by our beliefs in the particular expansion function and the particular density laws in question.[17] No matter how much evidence we had for the density laws and the extrapolation laws, it might be more reasonable to appeal to the causal principle and regard S_+ as a singularity whose causal antecedents cannot be described, rather than regard S_+ as a state without causal antecedents.

Even if we came to hold that S_+ was an uncaused state, the assertion that S_+ was the first event depends on further assumptions. To bring this out, suppose we have a theory which involves the above expansion and density laws and which also involves a function giving the mass of the universe as a function of time, such that the derivative of mass with respect to time is negative and constant. Extrapolating backwards we always have non-zero mass; extrapolating forwards we have, for some t, zero mass. If we suppose the metric tensor giving the space-time structure of this world collapses for zero mass, we seem to have reasons for thinking not just that there was an uncaused state S_+ but that there will also be a last state of the universe. This might look like evidence for the hypothesis that this universe had a first state S_+ and a last state S_n. However, the assertion that S_+ is uncaused undercuts any ground we might have had for denying that S_n will be followed by some uncaused event. That is, if we allow the violation of the causal principle we have no grounds for denying the existence of an uncaused state, perhaps rather like S_+, following S_n. Analogously, if we assert that S_n is the last state, we commit ourselves to giving up the principle of conservation of matter–energy, in which case we have undercut the grounds we might have had for denying that S_+ was preceded by state S_0, which, as it were, vanished without trace. In taking the uncaused state S_+ as the first state we are implicitly

[17] Indeed, there are physical theories according to which the transformations around the time of t_+ determined the current properties of matter. On this account of the matter we should see time t_+ as the limits of extrapolation on the basis of these laws. For before t_+ the constituents of the universe and their behaviour are governed by other laws. In this regard see the work of C. Hayashi cited by G. Gamov, 'Modern Cosmology', *Scientific American*, 190 (1954): 55–63: 63.

appealing to the principle that we have jettisoned in talking of the end of the universe.

This interplay between the causal principle and the principle of the conservation of matter–energy suggests that we will never be in a position to justifiably assert the existence of both a beginning and an ending of the universe. And, in relation to the earlier argument concerning a beginning universe, it brings out that we are implicitly appealing to a principle of conservation in passing from the claim that S_+ is uncaused to the claim that S_+ was the first state of the universe, the argument here being that any previous states would satisfy the principle of conservation and hence would not vanish without trace.

It is thus more reasonable to plead ignorance about the causal antecedents of some very dense state of the universe a long time ago than to assert that the state was an uncaused first state. However, if one is led to assert the existence of a first state on the basis of beliefs in the appropriate expansion, density, and conservation laws, we should deny the existence of earlier times. For if there are earlier times, there is clearly a violation of the conservation principle. For a state with zero matter–energy would be followed by a state with non-zero matter–energy. Thus the line of argument which assumes the non-violation of a conservation principle must take the time of the first state as the first time.

For reasons of the above sort it is difficult to envisage within our current scientific framework any viable theory that involves positing both a first event and time before that event. If such a postulation is not to be entirely idle, the theory would have to involve the claim that there must be some difference between a first-event, first-time world and a first-event, no first-time world (other than that difference)—a difference which the theory would account for by reference to time before the first event. This is difficult to envisage. In any event, it would clearly involve violations of the principle of the conservation of energy. Thus, it could not, for instance, come out of the field equations of the General Theory of Relativity as they have built-in conservation principles. And, as we remarked above, once one allows massive violations of the conservation of energy, one loses one of our most important methodological tools used in guiding our theory construction. Certainly one might envisage minor systematic violation of the principle of the conservation of energy. However, what would be involved in a first-event world with empty time prior to that event would be massive violations of conservation of energy. Hence the postulation of empty time prior to a first event is idle. Leibniz's remark that 'time without things, is nothing else but a mere idle possibility' *may* indicate that he held this view concerning time before a first event (i.e. that while there is no incoherence involved in talk of time before a first event, there would be no point in positing the existence of such time).[18]

The argument of the particular form that has been considered depends on the assumption that states of infinite density are physically impossible. This is, in fact, a contentious assumption. For it has become common to interpret the work on singularities in General Relativity as showing that the theory admits of the possibility of 'black holes', which are singularities where density is infinite.[19] In which case we are no longer licensed to postulate the state S_+ as the first state on the grounds that any state prior to S_+ would have a physically impossible density. It appears that the prospects for ever having evidence for a genuine first event are remote. For, supposing that the Big Bang emerged from a singularity of infinite density, it is hard to see what would constitute a reason for denying that that singularity itself emerged from some prior cosmological goings-on. And as we have reasons for supposing that macroscopic events have causal origins, we have reason to suppose that some prior state of the universe led to the production of this particular singularity. So the prospects for ever being warranted in positing a beginning of time are dim.

[18] *The Leibniz–Clarke Correspondence*, 75.
[19] See in this regard S. W Hawking and G. F. R. Ellis, *The Large Scale Structure of the Universe* (Cambridge: Cambridge University Press, 1973).

XI

TIME'S SQUARE

MURRAY MACBEATH

Two babies were born on the first day of this century, 1 January 1901. One
of them, Zoë, died exactly eighty years later, on 1 January 1981. The other,
Adrian, died, again on New Year's Day, in 1941. But, lest you think it sad
that Adrian should have enjoyed so much less of life than Zoë, let me assure
you that he packed just as much into his forty years as Zoë did into her eighty.
By the time Zoë was 5 months old and beginning to crawl, Adrian was
walking, uncertainly but rather fast; by the time Zoë was 13 years old and
reaching puberty, Adrian appeared to be in the prime of active (indeed,
hyperactive) young manhood; by the time Zoë was 30 and about to become
a mother for the third and last time, Adrian's hair was totally grey and, having
recently suffered a heart attack, he was climbing stairs not very much faster
than Zoë; and by the time Zoë was 40 and her hair was showing the first
streaks of grey, Adrian was wizened and, as I have told you, on his deathbed.
To look at and listen to Adrian was like looking at and listening to a film run
at twice normal speed. Clearly the fast-living Adrian was a biological freak.
But Adrian's (equally fast) friends would tell you otherwise. They noticed
nothing abnormal about his development: it is the events of Zoë's life that
they considered to have taken place at an unusual rate. You may expect me
to say that, to Adrian's friends, it appeared that the events of Zoë's life
unfolded unusually slowly. But no: they perceived the events of Zoë's life
as unfolding unusually fast. The view of Adrian's friends was that, by the
time Adrian was beginning to crawl, Zoë was walking, uncertainly but very
fast; that, by the time Adrian was reaching puberty, Zoë appeared to be in
the prime of active young womanhood; that, by the time Adrian was about
to become a father for the second and last time, Zoë's hair was quite grey
and she was finding the stairs tiring; and that, by the time Adrian's hair began

This essay has not previously been published. © 1993 Murray MacBeath.
 I am very grateful for comments made when I read earlier versions of this essay to the Moral
Sciences Club in Cambridge and to the Scots Philosophical Club in Stirling: especially helpful were
the comments of Peter Clark and Neil Tennant. That the diagrams are as clear as they are (in purely
graphic terms) is thanks to Denise Macrae and Ron Stewart of the University of Stirling's Media
Services.

to turn grey, a wizened Zoë was already on her deathbed. Adrian's friends would say that it was Zoë who looked like a speeded-up film, that it was Zoë who was the fast liver, the biological freak.

I shall have more to say about Adrian and Zoë later. But first the theoretical background. In his book *Space and Time* Richard Swinburne makes the following claim: 'Time, . . . being of logical necessity unique, one-dimensional, and infinite, has of logical necessity a unique topology.'[1] Swinburne attributes to time what I shall call the standard topology: he thinks that it is (1) unique—that is to say, there are not two or more temporally unrelated time systems; (2) open—as opposed, for example, to cyclic; (3) infinite—time has neither beginning nor end; (4) continuous—there are no 'gaps between periods of time';[2] and (5) one-dimensional. Swinburne also thinks that time has these topological or structural properties as a matter of logical necessity. Many philosophers believe that, if it is true that time has the standard topology—or, indeed, if it is true that it has some other topology—this is a contingent truth, not a matter of logical necessity. The way in which the contingency of propositions about the topology of time is argued for is very often the argument from fantasy, which describes an allegedly possible world in which there are, for example, two temporally unrelated time systems. In this case, in fact, the case of the attack on the claim that time is of logical necessity unique, it would seem that, for the possibility in question to be more than idle, more has to be established than just the possibility of a world containing two time systems: it has to be shown that there is a possible world with two temporally unrelated time systems, inhabitants of at least one of which have evidence for the existence of the other. Anthony Quinton, in his essay in this volume, in effect denies this possibility. Its most able defender, with his fantasy tale of the Okku and the Bokku,[3] was, ironically, the very Richard Swinburne who later recanted and argued time to be 'of logical necessity unique'. Those who believe, as Swinburne did, in the possibility of a two-time world may hold that our world is, for all we know, such a possible world; that is to say, for all we know, time actually is not unique.

A good deal of attention has been devoted to the first four properties that I listed in characterizing the standard topology. Far less has been devoted to the fifth, to the dimensionality of time. Bill Newton-Smith, for example, in

---

[1] Richard Swinburne, *Space and Time* (London: Macmillan, 1968), 209.

[2] Ibid. 208. In the second edition (1981), Swinburne removes from the topology chapter the paragraph containing these words, and with it his claim that time is 'of logical necessity . . continuous'; and when he does touch on the question, in the chapter 'Past and Future', he appears willing to countenance the possibility that time is discrete.

[3] Richard Swinburne, 'Times', *Analysis*, 25 (1965): 185–91.

his book *The Structure of Time*,[4] devotes four of his ten chapters to the topology of time (one chapter for each of the first four properties in my list), but does not mention dimensionality. In a way, the neglect of the dimensionality of time is odd, for there has been much discussion of the dimensionality of space. And indeed, since the question of the extent of the similarity between time and space is constantly discussed, and since dimensionality, along with directionality, is one of the two generally agreed dissimilarities between time and space, it is even more odd that the dimensionality of time is so infrequently investigated. Seen from another angle, however, the neglect is not surprising. In giving a brief and rough account of what is involved in the attribution to time of the standard topology, I was able, in the case of the first four properties, to say what is being denied when it is claimed that time is unique, open, infinite, and continuous. But what is being denied when it is claimed that time is one-dimensional? That it is two-dimensional, of course; but what on earth (or elsewhere) could two-dimensional time be like? Swinburne's arguments notwithstanding, we can, I think, get hold of the idea of time as not open but cyclic. But the idea of time as two-dimensional is, initially at least, quite baffling. We need, of course, some account of what a dimension is. A first shot, and a familiar account (associated with the name of Riemann), is this: in saying that a space has n dimensions, we are saying that the location of any point in that space requires that values be specified for n parameters. The most obvious kind of example is that of indicating, by reference to a map, the location occupied by a geographical feature. Here we say how far to the east or west the feature is of a given point of origin, how far north or south it is of the same (or some other) point, and how far it is above or below sea-level. This case is based on spherical geometry, and the coordinates are therefore not Cartesian. But this does not affect the assertion that, for the location of any point in physical space, we have to supply a point of origin and values for three parameters, and that this is what is meant by the claim that physical space is three-dimensional. Unfortunately, this first shot will not do. For, as Cantor first showed in 1877, there are as many points on a line as in a plane, so that the two sets of points can be paired off one-to-one. It is then possible to locate any point in the plane by citing the single coordinate of its corresponding point on the line. Our familiar three-dimensional space would turn out to be one-dimensional if we accepted the first shot at a definition of dimension.

A more recent and technical definition of dimension is in terms of 'cuts'. A line, which is a continuum, can be cut into two separate continua by the removal of a point, which is not a continuum: a line is therefore

[4] London: Routledge & Kegan Paul, 1980.

one-dimensional. The definition then proceeds recursively: a continuum is n-dimensional, where n is an integer greater than one, if it can be cut by the removal of a continuum of dimension $n - 1$. Thus a square can be cut into two rectangles by the removal of a line, or a sphere into two hemispheres by the removal of a circle. The notion of a cut can be applied, much less rigorously, in the following way: that we are spatially three-dimensional is shown by the fact that, whereas, if I am standing in a (two-dimensional) ring, I can step out of it without cutting the ring, if I am enclosed in a (three-dimensional) sphere, I cannot get out without cutting it. This second account of dimension is not an easy one to work with when discussing the dimensionality of time. It may also seem, perhaps surprisingly, to have little to do with the more familiar first account. It will be helpful, therefore, to investigate the connection between them. Now if we draw a continuous curve in a plane, its Cartesian x-coordinates and y-coordinates will each themselves form a continuum. However, as was established by Brouwer,[5] this is not true of the single Cantor coordinates for the curve that are produced by mapping the plane on to a line. That is to say, Cantor's transformation destroys the topology of the space. Hence our first shot will after all suffice if, to quote van Fraassen, 'we insist that the assignment of coordinates reflect the topological properties of the space'.[6] Now, as van Fraassen goes on to say, 'if dimension is a topological invariant, then the detour via coordinates is superfluous and dimensionality should be defined in topological terms'[7]—hence Poincaré's definition in terms of cuts. However, for the sake of clarity—because the notion of a cut is a technical one, and its application to the dimensionality of time is at least not straightforward—I suggest that we look for a more accessible, even if less rigorous, account of dimensionality, one that does justice to Cantor's discovery and exploits Brouwer's.

Such an account is, I think, provided by saying that a space has n dimensions if there are n respects in which its occupants can, *qua* occupants of that space, vary continuously but independently. This is a good deal less formal than the classical definitions, and it is only offered as a working model; but it does accommodate, for example, the claim that colour is a three-dimensional space, in that colours can vary (or objects can vary in colour) continuously but independently in hue, saturation, and brightness. We might be inclined initially to say that, for the location of any particular colour in this logical space, we need to provide values for these three parameters, since

[5] L. E. J. Brouwer, 'Beweis der Invarianz der Dimensionenzahl', *Mathematische Annalen*, 70 (1911): 161–5.
[6] Bas C. van Fraassen, *An Introduction to the Philosophy of Time and Space* (New York: Random House, 1970), 134.
[7] Ibid.

a red can have the same saturation and brightness as a blue, or a green the same hue and brightness as a red. But decorators' colour charts do, Cantor-like, relate particular colours to single numbers. However, even if an infinite colour chart allocated single real numbers to every particular colour, the continuous range of colours displayed by, say, a ripening tomato would not be matched by a continuous range of numbers. The fact that my proposed definition allows us to talk of colour as three-dimensional ensures that we are not working with an account of dimensionality which relies on spatial vocabulary or imagery. That being so, it may be easier for me, in giving an account of two-dimensional time, to rebut the charge that I have spatialized time. A world in which time is two-dimensional is a world in which the temporal properties of events can vary continuously in two independent ways. If I subsequently represent such independent variation in the form of diagrams, this no more shows that I am spatializing time than would the claim that colour is three-dimensional, supported by diagrams, represent a spatia-lizing of colour.

At this point Adrian and Zoë make their reappearance. I shall be concerned to examine the claim that their story, which I told at the outset, requires the postulation of two-dimensional time. Their story is my reworking of a tale originally told by Judith Jarvis Thomson.[8] I think her account is still the best that has been given of what circumstances might force upon us the hypothesis of two-dimensional time;[9] but it involves at least three connected mistakes, which I shall try to put right. Why should the story of Adrian and Zoë require the postulation of two-dimensional time? Intuitively, the idea is this. We are initially inclined to describe Adrian as having lived half as long as Zoë but at twice the rate. Then we discover that Adrian's friends, who are just as numerous and just as sane as us, are inclined to describe Zoë in the way that we describe Adrian. Does this not suggest that perhaps the apparent dif-ference between their life-spans and life-rates is to be accounted for by a difference in perspective? We have no reason to think that Adrian's life was shorter than Zoë's that cannot be matched by a reason for thinking that Zoë's life was shorter than Adrian's. Did each live half as long as the other, by different measures? Or could their lives have been of the same length, by some unknown measure? If we try to draw the lifelines of Adrian and Zoë in such a way that both are straight lines of equal length, but also that coheres with our observation that Zoë lived from 1901 to 1981 while Adrian lived

[8] Judith Jarvis Thomson, 'Time, Space and Objects', *Mind*, 74 (1965): 1–27, at 20–7.
[9] Other philosophical accounts of two-dimensional time include Jack W. Meiland, 'A Two-Dimensional Passage Model of Time for Time Travel', *Philosophical Studies*, 26 (1974): 153–73, and T. E. Wilkerson, 'Time and Time Again', *Philosophy*, 48 (1973): 173–7, and 'More Time and Time Again', *Philosophy*, 54 (1979): 110–12.

from 1901 to 1941, the only way in which this can be done is by drawing Adrian's lifeline at an angle to Zoë's. The vertical axis represents time as we ordinarily conceive it—or, as we may now have to say, the one dimension of time with which we have hitherto been familiar. There is a horizontal axis drawn at right angles to the vertical, if only because that is the Cartesian thing to do. But, while Zoë's lifeline runs parallel to the vertical axis, Adrian's lifeline is at an angle of 60° to Zoë's lifeline and to the vertical axis, as in Figure 1.

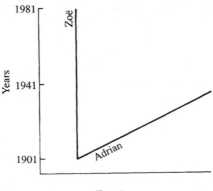

FIG. 1

This intuitive approach, of course, does involve just the kind of spatializing of time that I said I wished to avoid. For it asks, in effect, what the spatial properties of a given time-diagram would have to be. But the diagram can be justified independently of this intuitive approach. What the diagram does is to represent the manner in which events vary continuously in two independent ways. Adrian sees both Zoë's life and his own as continua, hers half as long as his, while Zoë sees both Adrian's life and her own as continua, his half as long as hers; and the independence of these two viewpoints makes it impossible for us to accept both their observations and to place both their lives in a one-dimensional continuum of time. When shown the diagram I have presented in Figure 1, Adrian's friends will not disagree with the relationship between the two lifelines. However, they will say: 'Time as *we* ordinarily conceive it is not represented by the vertical axis you have drawn. A line that properly represented the time with which we are familiar would be parallel to *Adrian*'s lifeline. And we measure time, not in years—as you assumed in your exposition—but in yonks. However, as the earth revolves around the sun twice a yonk, we reckon that the lapse of a yonk for us is the same as the lapse of one of your years for you. Adrian, incidentally, was born

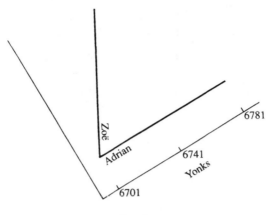

Fig. 2

in the yonk 6701 and died in 6781, so you won't be surprised to hear that our records show that Zoë, who was also born in 6701, died in 6741.' The reworking of Figure 1 that Adrian's friends will propose is Figure 2.

Now if we twist and rotate this diagram in order to give it a more conventional orientation, as in Figure 3, we find that it comes to bear a striking resemblance to the original diagram (Figure 1), a resemblance which reflects the epistemological symmetry that is an important feature of the story of Adrian and Zoë. It is, of course, a crucial aspect of this story that we have two eye-views of the situation. If we had only the perspective of Zoë's friends, we would simply conclude that Adrian was a fast liver. It is the testimony of Adrian's friends, who have the same view of Zoë as Zoë's friends do of Adrian, that seems to push us in the direction of postulating a second time dimension. However, it might be objected that such testimony is in principle unavailable to us, and that a case of this kind for the two-dimensionality of time must therefore lack an essential ingredient. This objection I derive from something that Judith Jarvis Thomson herself says about people like Adrian: 'we should . . . have to *say* that it is absolutely impossible really to communicate with them'.[10] If indeed communication between, on the one hand, Zoë and her friends, and, on the other, Adrian and his friends, is absolutely impossible, then the one group can never know how the other perceives it, and neither side will have any reason to believe that the other side consists of anything but fast livers. I could simply argue that,

[10] Thomson, 'Time, Space and Objects', 26.

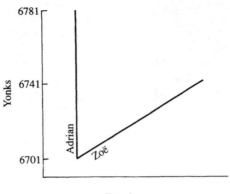

FIG. 3

even if it were impossible for anyone to know that both an Adrian-perspective and a Zoë-perspective were realized, none the less it might be the case
that both were realized. I shall be bolder, however, and argue that, on a
plausible-looking set of assumptions, communication between Adrian and
Zoë is possible. And I hope, therefore, to be able, relative to those assumptions, to defend the view, not just that there could exist an Adrian whose
lifeline was skewed with respect to Zoë's and ours, but that Zoë and we
could have reason to believe that this was so. First, then, let us look at
Thomson's reasons for thinking that communication between Adrians and
Zoës is absolutely impossible. She argues:

> If I say or do something to one of these people, then even if it should seem to me that
> he reacts to this, we should have to say: though he appears to be reacting now to what
> I just said or did, he is not reacting to this, but rather to something which will not be
> said or done for many years to come. For my sayings and doings are not really
> contemporaneous with the events in his life which seem to be contemporaneous with
> them.[11]

This argument needs some clarification. The point is, I take it, not, as the
second sentence might suggest, that the events of the sending and receiving
of a message must be (very nearly) contemporaneous if the process is to be
one of communication. If that were the point, it would be clearly wrong, for
I can communicate by surface mail with my brother in Winnipeg. And
sufficiently long-lived and technologically sophisticated creatures could
communicate with each other even if they lived in galaxies that were thou-
sands of light-years apart. The point is, rather, that the facts which lead us

[11] Ibid.

to speak of an angle of 60° between Adrian's and Zoë's lifelines also generate insuperable temporal dislocations in any attempted communication between them, dislocations that take the form not just of long delays but of anticipations. Consider Zoë at the age of 32 (years), at point X (in Figure 4) on her lifeline. She perceives Adrian as ageing at twice her own rate, so to her Adrian appears to be 64 years old. She decides to try to communicate with him, and asks 'Where do you get all this energy from?' Adrian, as it happens, hears the question, but not when he is 64 (yonks), not, that is, at point Y on his lifeline. Rather, since he perceives Zoë as ageing at twice his own rate, and since the question is asked by a woman who is apparently 32 yonks old, he hears the question asked when he is 16 (yonks), at point W on his lifeline. He answers 'You're the one with the excess energy.' Zoë, as it happens, hears the reply, but not when she is 32 (years), not at point X. Rather, since she perceives Adrian as ageing at twice her own rate, and since the reply is uttered by a boy who is apparently 16 years old, she hears the reply when she is 8 (years), that is, at point V on her lifeline. Since the question to which she has now (at point V) heard the reply will not occur to her or be asked for another twenty-four years (until point X), she does not take this utterance of Adrian's to be a reply to her question.

Communication has not taken place. And since a reply will always be received, if at all, before the question to which it is a reply has been asked, communication is impossible. So, at least, it might be argued. However, I shall criticize this argument on two grounds, in two stages. The thesis of

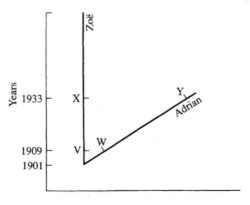

Fig. 4

stage one: communication of a kind has taken place; and even if communication of this kind is still problematic, there is an unproblematic minimal form of communication which is sufficient to enable Adrian to communicate his perspective to Zoë. The thesis of stage two: with a little bit of ingenuity, full-blown communication can be arranged between Adrian and Zoë.

Stage one. In the sequence of events just described, communication of a kind has taken place in that Zoë has asked Adrian a question and Adrian has heard it asked and has taken it to be addressed to himself. It is, no doubt, fortuitous, or a matter of lucky chance, that Adrian did hear the question: for this to be the case, he must have been in the same vicinity on the relevant day when he was 16 yonks old (point W) as he was to be forty-eight yonks later (point Y), when he was there to be seen by Zoë—the sighting that prompted her decision to try to communicate with him. It is also, no doubt, fortuitous, or a lucky chance, that he took the question to be addressed to himself, especially as he would probably be used to a complete lack of fit between his own behaviour and that of Zoë and her friends. These considerations may, of course, suggest to some that we should withhold the name 'communication' from the sequence of events that I have described, on the grounds that the requisite recognition of intentions is lacking.

However, in this stage one argument, I wish to focus on the question whether Zoë, as a result of an utterance (or inscription) of Adrian's, can come to know of his perspective on her life; so the question of the intentionality of the concept of communication can be left on one side. In stage two I shall attempt to show that even the richest concept of communication can be instantiated by a dialogue between Adrian and Zoë. Now an objector may still feel that the doubly fortuitous sequence of events described above is, even when the question of intentions is set aside, too far-fetched to be worth considering. So let us consider a case which does not have these fortuitous elements, a case in which Zoë simply says something and Adrian hears it, or in that Adrian writes something down and Zoë reads it. Now this minimal communication is all that is needed if Adrian's perspective on Zoë is to be communicated to Zoë, or vice versa. It could be effected by Zoë's hearing various things that Adrian says about her and her friends; but, because the supposition raises interesting questions, I shall ask you to suppose that Adrian observes Zoë closely over many yonks, and writes down his observations in a diary; and to suppose that Zoë finds the diary and is able to read it. She now has had communicated to her the Adrian-perspective on her own life; and, to her surprise, this perspective is not that she seems to the fast-living Adrian to be a slow liver, but that she seems to him, just as he seems to her, to be a fast liver. That the contents of Adrian's diary constitute a genuine perspective on a shared world, but a perspective that cannot be accounted for without a change in our understanding of the nature of time, might be confirmed by Zoë's discovery in the diary of a detailed narration of events that, as far as she is concerned, are in the future, but that later take place exactly as described. However, even the supposition of a diary, written by Adrian and found by Zoë, might be questioned. Ronald E. Nusenoff,[12] for example, criticizes Thomson on the grounds that any attempt to fit the physical objects with which we are familiar into the world that she describes yields results that are barely conceivable. His argument invokes a variant of Figure 5, differing from it, chiefly, in that Nusenoff, following Thomson, sets the angle between the lifelines at 45° (I prefer an angle of 60° since it coheres with observations of Adrian as living twice as fast as us, rather than √2 times

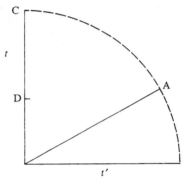

Fig. 5

as fast). I should also point out that Thomson and Nusenoff bestow upon Adrian the unlikely surname of Smith. Before developing his argument, Nusenoff quotes Thomson's claim that communication with people like Smith would be absolutely impossible, and he comments:

What this seems to mean is that we could not hear anything that Smith would appear to be saying. Nor, I suppose, could we feel him. Nor smell him. Why, then, should the sense of sight be so privileged so as to allow us to have visual contact with Smith? There seems to be no justification for this, so it must be admitted that if all

---

[12] Ronald E. Nusenoff, 'Two-Dimensional Time', *Philosophical Studies*, 29 (1976): 337–41. Nusenoff is the would-be scourge of the two-dimensionalists; see also his article 'Spatialized Time Again' (*Philosophy*, 52 (1977): 100–1), in which he argues that Wilkerson, in 'Time and Time Again', illicitly spatializes time. He would also say the same, no doubt, of the present essay. But it is worth saying, in passing, that my objection to the diagram in Wilkerson's article is that it temporalizes space. As is made abundantly clear by his fleshing-out of the diagram in his 'More Time and Time Again', the simplest interpretation of his diagram, so simple indeed as to be wholly unproblematic, is that the horizontal axis, which he calls the A-time axis, is a spatial axis.

forms of nonvisual perception are 'absolutely impossible', so must be visual perception.[13]

This is clearly a misinterpretation of Thomson, for she explicitly says that Smith reacts to what we say (before we say it, of course), which implies that he hears what we say; and, since we are in the same position with regard to him as he is with regard to us, Thomson's view must be that we *can* hear what he says. The bar to communication, in Thomson's view, is not that we cannot hear Smith, but that our hearing him is temporally dislocated from his speaking. Setting that point aside for the moment, we come to Nusenoff's argument concerning physical objects:

> Thomson's explanation of Smith's apparently fast life-rate is that he is travelling in a different temporal direction than we are. It is assumed that he is a member of our world, and so lives with the same physical objects as we do. Now suppose that we are at point D. If we could observe Smith, we would see him at point A. Let us further suppose that Smith is in the same room at point A as we are at point D. One question we could ask here is, 'What do we, at point D, see if Smith, at point A, picks up some object which is in the room at both points D and A?' The answer we must give is, 'Nothing!'
>
> Smith's picking up of the object is not contemporaneous with our presence in the room at point D. Even if we allowed visual contact with Smith himself, all we might see is Smith grasping the object, and then moving his hands through it as if he were picking it up. If we were to return to the room at point C, which is contemporaneous with point A in terms of real time, then we would see the object move. There would be no physical explanation for this movement, for even if, once again, we allowed visual contact with Smith, we would not observe him picking up the object, for we could only have observed his actions of point A from point D.[14]

I think that there are two important mistakes in this argument. A clue to the first mistake is provided by Nusenoff's stipulation that the object which Smith picks up—a diary, let us say—is in the room at both points D and A. Clearly it has to be in the room at point A, for Smith picks it up at point A, and he is then in the room. But why need the diary be in the room at point D? The answer one is inclined to give, and that Nusenoff would presumably give, is that we are in the room at point D, and at point D we see Smith pick up the diary. However, Smith is not at point D—his lifeline does not pass through that point—yet he is seen by us, who are at point D, to be in the room. So the diary could, like Smith himself, be seen by us, who are at point D, without itself being at point D. Now Nusenoff would hold that, in the last passage that I quoted, the argument of his second paragraph proceeds *per impossibile*, based, that is, on the false assumption that we could see Smith. However, no good reason has yet been given why we should not be able to see Smith—the only reason Nusenoff has given was based on a misinterpretation of Thomson. So I conclude that, since we, at point D, could see Smith

---

[13] Nusenoff, 'Two-Dimensional Time', 340.   [14] Ibid.

without his being at point D, we could see the diary without its being at point D. The mistake that Nusenoff has made is to overlook the fact that a physical object will, like Smith himself, have its own lifeline, and that its observable behaviour will depend on the orientation of its lifeline relative to that of the observer, in precisely the same way as the observable behaviour of a person such as Smith depends on the orientation of his lifeline relative to that of the observer. Nusenoff's second mistake he inherits from Thomson. She said, remember, that 'my sayings and doings are not really contemporaneous with the events in Smith's life which seem to be contemporaneous with them'. In saying this she was developing an earlier comment about the deaths of Smith and herself—in her article Thomson herself plays the role that I have allotted to Zoë:

[Smith's] death appears to me to precede mine, and mine appears to him to precede his—neither preceding the other, but both in fact occurring in the same amount of real time—80 real years—after our births. . . . We shall have to say this about these people: the events we seem to be observing in their lives we are not really observing, for the events we seem to observe are not contemporaneous with our observings.[15]

What does Thomson mean by '80 real years'? If a real year is the interval during which the earth revolves once around the sun, then whereas Thomson's death occurs eighty real years after her birth, Smith's death occurs, not eighty, but forty real years after his birth. (This figure of forty again relates to a diagram in which there is a 60° angle between the two lifelines: Thomson's own diagram has an angle of 45°.) Now imagine Max, who is born at the

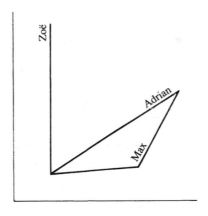

FIG. 6

[15] Thomson, 'Time, Space and Objects', 25–6.

same time as Adrian Smith, Judith Jarvis Thomson, and Zoë, but who differs from them in that his lifeline, as drawn from our perspective, is bent: until he looks like, and has achieved what might be expected of, a man of 45, he seems to us to age even faster than Adrian; thereafter, until his death, when he looks like a 90-year-old, he seems to us to age more slowly than Adrian, but still faster than Thomson, Zoë, and ourselves. And Max appears to us (and, incidentally, to Adrian's friends) to die at the same time as Adrian. (See Figure 6.) Does Max's death occur eighty real years after his birth, or ninety real years after his birth? Because it is not clear how this question is to be answered, I think that, in the present context, the notions of 'real years' and, more generally, of 'real time' are not helpful ones. I also think that, if the notion of contemporaneity (or, slightly less polysyllabically, simultaneity) is to be retained, it will have to be relativized to the conception of time as two-dimensional. Rather than speaking of events as simply simultaneous, we should speak of events as simultaneous with respect to a certain dimension. Because any person's experience still consists of a linearly ordered sequence of events, Zoë will, as we have already seen, perceive as *simply* simultaneous events at X and Y in Figure 4, while Adrian will perceive as simply simultaneous events at W and X; and we can add that someone whose lifeline ran parallel to the x-axis would perceive events at V and X as simply simultaneous. But, trying to fit their different experiences into a unified picture, we have had to resort to a two-dimensional structure. So, rather than ask 'Which of these pairs of events are really simultaneous?', I think we should say that X and Y are y-simultaneous, whereas V and X are x-simultaneous. In Figure 7 (on p. 198) the labels that were used in Figure 4—'V', 'W', 'X', and 'Y'—have been replaced with 'E', 'F', 'G', and 'H' respectively. So, using the Figure 7 labels, I think we should say that G and H are y-simultaneous, whereas E and G are x-simultaneous.

Now let us return to Nusenoff and to his claims about what we would see at points D and C. He claims that at point D, if we were to see Smith in the act of picking up the diary, we would see his hands pass through it; and he claims that at point C, if we saw anything, we would see the diary move without any corresponding movement on Smith's part. (Indeed, at point C Smith has already been dead for forty years.) Why does Nusenoff think that our observations would take this form? Because it is 'point C, [and not point D,] which is contemporaneous with point A [the point at which the diary is picked up] in terms of real time'.[16] But events that we perceive as simply simultaneous are events that are y-simultaneous. Now the motion of Smith's hands in picking up the diary, and the motion of the diary as he picks it up,

[16] Nusenoff, 'Two-Dimensional Time', 340.

are surely y-simultaneous; so at point D we will observe them as occurring together. Thus we will see Smith picking up the diary and the diary being picked up by Smith in the normal way, though at twice normal speed. What we will see at point C is a quite distinct set of events, events that are y-simultaneous with point C. I might add, finally, that, even if the observation of events as occurring together did depend on 'real time' simultaneity, then Smith's movement and the movement of the diary would still be observed together, for both take place at point A and both would thus be observed at point C. Nusenoff is able to divorce Smith's movement from the movement of the diary, and thus make the story of observing Smith's interaction with a physical object seem 'barely conceivable', only by assuming that the observation of Smith's behaviour depends on y-simultaneity, while the observation of the behaviour of the diary depends on 'real time' simultaneity. Nusenoff's second mistake, that of using in a confused way the notions of 'real time' and of events that are 'really contemporaneous', here intersects with his first mistake, that of failing to see that what goes for persons goes also for physical objects. The argument just concluded was designed to meet an objection to the claim I made earlier, that Adrian's perspective on Zoë could be communicated to Zoë by her finding and reading Adrian's diary. The objection was that physical objects, such as a diary, cannot coherently be fitted into the story of Adrian and Zoë; and I have tried to meet the objection by pointing out mistakes in the argument advanced in its support by Nusenoff. There is a closely related issue to which I shall have to return, but for the moment I will move on to the second stage of my larger argument about communication between Adrian and Zoë.

Stage two. Despite the temporal dislocation between a question asked by Zoë and a reply offered by Adrian, full-blown communication between them is not, as Thomson thinks, impossible. We have seen that, if Zoë asks a question of Adrian when she is 32 years old, and, if he hears the question, takes it to be addressed to himself, and answers it immediately, then, if Zoë hears the reply, she will hear it when she is 8 years old. Full-blown communication could not take place between people so oriented to one another in time unless it were possible for full-blown communication to take place between us and someone living in reversed time. Now it is usually held, and it is forcefully argued by J. R. Lucas[17] in particular, that full-blown communication with people living in reversed time is impossible. I have tried elsewhere[18] to show that Lucas is wrong; and I can adapt the method that I there proposed for 'time-antagonists' to our present case of 'time-divergers',

---

[17] J. R. Lucas, *A Treatise on Time and Space* (London: Methuen, 1973), 45–6.
[18] Murray MacBeath, 'Communication and Time Reversal', *Synthese*, 56 (1983): 27–46.

Adrian and Zoë. Let us again assume that Zoë, 32 years old, looking at Adrian, whose apparent age is 64, decides to try to communicate with him. She has a question to ask him, and she wants to hear his reply to it soon after she has asked the question. Now she has reason to believe that, if she transmits her question immediately, that is, at point G, any immediate reply that Adrian gives will not be heard now, but will have been heard, if at all, twenty-four years ago, at point E. However, by the same token she has reason to believe that there is a temporal point I such that, if she delays the transmission of her question until point I, Adrian will receive the question at point H, and that, if he answers the question very soon after point H, she will hear his answer very soon after point G, which is exactly when she wants to hear his answer. Point I is most plausibly located on Zoë's lifeline or a projection of that line into the y-future beyond her death. Point I is also on a line drawn through point H at right angles to Adrian's lifeline; for it is events on this line that Adrian perceives as being simply simultaneous with point H, and he is to receive at point H the question that Zoë transmits, or arranges to have transmitted, at point I. Point I therefore lies in the year in which Zoë, had she survived, would have been 128 years old.

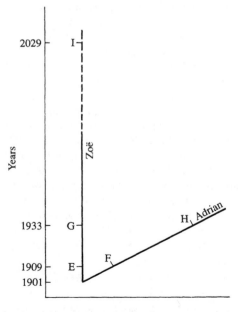

Fig. 7

We are now in a position to see how communication between Adrian and Zoë can take place. Zoë writes down her first message, and arranges that it will be displayed in the room where she now sees Adrian, but not until after a ninety-six-year delay. There are several ways in which she might arrange this delay in the transmission of her message: she might leave instructions in her will; she might install in the room a computer that can be programmed to store a message and display it after a specified interval; she might leave an envelope on a table, with instructions for when it is to be opened. Any of these methods might fail, in which case Adrian will not receive Zoë's message, or will perhaps receive it at a point earlier or later than point H. The attempt to communicate across two-dimensional time is an enterprise with many risks of failure; but the fact that these risks exist in no way supports Thomson's view that such communication is 'absolutely impossible'. I said that I intended to defend the claim that, given certain plausible assumptions, 'full-blown' communication was possible between Adrian and Zoë, by which I meant a series of messages, each of which is a response to the previous one, and where there is no arbitrary limit to the length of the series. I still need to show, therefore, that the dialogue between Adrian and Zoë can be prolonged after her initial question and his reply. This is easily done. Having received Adrian's reply, Zoë can frame a new message, and arrange for it to be displayed at a suitable interval after the eventual display of her first message. If the interval is four hours, and if Adrian replies to this question, as to its predecessor, immediately, then Zoë will hear his second reply one hour after she heard his first. She can then frame her third message and decide on a time for its display. And so the process can go on, until it is terminated by accident, mistake, boredom, or death.

I said earlier that Judith Jarvis Thomson's account of two-dimensional time involved at least three connected mistakes. I have so far pointed out two. The first was her talk of 'real time' and of events as 'really contemporaneous'. (Since she also talks of people 'moving to the right or left in time',[19] maybe she would be prepared to tolerate the idea of time itself as moving, in which case her real time would move, not like an ever-rolling stream, but like an ever-expanding ripple, which reaches the ends of Adrian's and Zoë's lifelines at the same real instant.[20] Quite apart from the dangers

---

[19] Thomson, 'Time, Space and Objects', 27.

[20] In her diagram (which is not importantly different from my Figure 5) Thomson makes the geometric origin (the point of intersection of the x- and y-axes) coincident with the point at which she and Smith are born. Because it is surely presumptuous to suppose that real two-dimensional time should have as the source (or metaphysical origin) of its outward rippling motion the moment of one's own birth, and arbitrary to suppose that it ripples out from *anyone*'s birth, I have chosen to represent Zoë's lifeline as parallel to but not coincident with the y-axis, and the geometric origin (which has no metaphysical significance as the source of time-ripples) is at an arbitrary point before Zoë's birth.

involved in talking of people as moving in time, or of time itself as moving, I think, and have argued, that Thomson's talk of real time has no clear sense or value.) The second mistake that I claim to have detected is her assertion that communication between people such as Adrian and Zoë would be absolutely impossible. To locate the third mistake we have to go back to Nusenoff's objection that physical objects cannot be fitted into the story of Adrian and Zoë. I argued that that objection could be met, since Nusenoff had overlooked the fact that physical objects have their own lifelines. Thus how a physical object is seen to behave—for example, when causally acted on by a person—depends on the orientation of its lifeline to that of the observer, just as how the person acting on it is seen to behave depends on the orientation of his or her lifeline to that of the observer. But there is a price to pay for saving Thomson from Nusenoff's objection, and it is this. If physical objects have their own lifelines, we can speak of Zoë and Adrian, as I did recently, as being in the same room, she at point G on her lifeline, he at point H on his, only if the room itself has a lifeline that passes through points G and H. To accommodate them both at these two different temporal instants, the room will have to have an interesting history of its own, quite apart from the history made by its occupants. For any lifeline that we draw for the room passing through both points G and H either (1) will be parallel to the x-axis, in which case all the events in a finite period of the room's history will seem to Zoë to be simultaneous, or else (2) will double back on itself with respect to the y-axis, in which case for part of its history Zoë will see events in the room's history as running in reverse. There are, then, deeper reasons than Nusenoff himself mentions for finding it difficult to imagine this two-dimensional time system as inhabited by both people and physical objects, if, that is, the time-diverging people and the physical objects are all in the same area of space. However, if we put Zoë and Adrian in different places, in surroundings constituted by different physical objects, this difficulty can be significantly reduced. I propose, therefore, that we regard Adrian and Zoë as inhabitants of different planets. There is, in any case, good reason for doing this, since an Adrian suddenly appearing in this room would play merry hell, not only with the furniture, but with our physics. And I think it important, in putting forward a fantasy argument of the kind I have been considering, to retain as much of our physics as can be retained; that is, in producing a description of a possible world, to ensure that it is governed by scientific laws maximally similar to the scientific laws that govern the actual world. If we put Adrian and Zoë on different planets, or better in different galaxies, it might be thought that the only alterations that will need to be made to our physics will be at what is anyway one of its most fluid points, in astrophysics. Our ordinary terrestrial physics would continue to hold good on earth, and it would hold good on Adrian's planet too, except that time-dependent laws would there apply relative to a skewed time-scale. Separating Adrian and Zoë in this way would remove the problem of integrating their different behaviour patterns into a shared world of physical objects, and it might be hoped that any boundary problems that arise would prove soluble.[21]

Unfortunately, this will not do. I said earlier that, given certain plausible-looking assumptions, there was no reason why full-blown communication between Adrian and Zoë should not be possible. The qualification stands in need of explanation. One assumption that I made, and that does look plausible, is that, if two people are y-simultaneously close together in space, with no other objects between them, then the person whose lifeline is parallel to the y-axis will be in a position to see the other person, whatever the relative orientation of their lifelines. This assumption does seem to be made by Thomson; and, though it is rejected by Nusenoff, it is, as I have earlier pointed out, rejected by him solely on the basis of an *ad feminam* argument which is itself based on a misinterpretation of Thomson. However, the assumption is worth examining more closely. For what determines whether or not I can see another person at a given point in time is, less where he or she is relative to me in space-time, than whether there is, connecting his or her lifeline and mine, the lifeline of an appropriate cluster of photons.[22] Unless such a cluster of photons travels in space between that person and me, then, whatever our orientation in time, I cannot see him or her. So we must ask: can a photon lifeline be drawn in such a way as to connect, for example, points G and H in Figure 7, so that Zoë at point G will be able to see Adrian at point H? Whether it can or not is a question to which there is no straightforward answer. For if photon lifelines are to link the diverging lifelines of two people, then to both of these people the photons will appear to travel faster than the speed of light. So our physics is going to have to be altered. Further, given that lifelines are directional, are we to allow photon lifelines to point in any direction across two-dimensional time? It used to be said that if an American stood in Times Square, New York, he would eventually see everyone he knew. Can we arrange our physics so that people like Adrian and Zoë will be able to see each other, as they follow their different routes

---

[21] If this inter-galacticization of Adrian and Zoë seems to you to be viciously *ad hoc*, I would ask you to consider that I could have begun this essay quite differently. Imagine that it started along the following lines: Zoë the eminent astronomer observes that in the galaxy Citius physical processes seem to take half the time they do in our Milky Way galaxy; when she later receives messages which make it clear that there is intelligent life in Citius, she is astonished to hear from the Citian astronomer Adrian that he thinks that physical processes in our galaxy take half the time they do in Citius.

[22] The same, of course, goes for hearing someone and sound-waves; and this observation must qualify all I have said about Zoë and Adrian as hearing one another.

across the time's square that I have drawn? Probably. But we are now in a position to see that no determinate answer can be given to the question what it would be like to live in a two-dimensional time system of the kind that Thomson thinks her story would require us to postulate. There is no single thing that it would be like, for all would depend on the physics that operated; and there is no single obvious way in which our physics can be adapted to fit a world with two-dimensional time. For example, would photons emitted from a light source have lifelines that were extensions in a straight line of the source's own lifeline; or would the photon lifelines spread out in all temporal directions? Would each person be able to alter his or her own orientation in time, and to alter that of physical objects and other people? Would the temporal properties of things be such that their lifelines should be represented as having width as well as length? Even lacking the answers to such questions, I do agree with Thomson that there are certain observations that we might make which would be sufficient to support the thesis that we do in fact live in two-dimensional time; and, of course, the observations might actually supply the answers to the questions I have just raised, telling us something about the operative physics. But perhaps I should be more cautious, and say that there are possible observations which would support the thesis that time is at least two-dimensional; for I would not want to rule out the possibility of further observations which supported the thesis that time is three-dimensional. Or worse.

XII

SPACES AND TIMES

ANTHONY QUINTON

We are accustomed to thinking of space and time as particulars or individuals—even if we should hesitate to describe them as things or objects or substances. We say 'Space has three dimensions', 'Material things occupy space', 'The debris has disappeared into space', and we talk in a comparable fashion about time. Not only do we think of space and time as individuals but, in many connections at any rate, we think of them as *unique* individuals. When we talk about spaces and times in the plural, when we say 'Fill up the spaces on the form', 'It could go in the space between the lamp and the door', 'There were peaceful times in the early years of their marriage', we think of these multiple spaces and times as parts of the unique all-encompassing space and the unique all-encompassing time. Kant believed that we could not help thinking of them in this way. We do, at any rate, in fact think like this and it is this conviction that I want to examine. What, I shall ask first of all, does the belief that space and time are unique individuals come to? Secondly, is the belief in either case true? Finally, if it is true in either case, is it necessarily true, or is it simply a matter of fact?

1

What, to start with, does it mean to say that space is a unique individual? We could say instead that all real things are contained in one and the same space. Two things are in the same space if they are spatially connected, if there is a route connecting them, if each lies at some definite distance and in some definite direction from the other. The relation of spatial connection is clearly symmetrical. If I know the route leading from A to B, I must also know the route leading from B to A. It is also transitive. Given the route from A to B and the route from B to C the route from A to C is unequivocally determined. Now it does not follow from these properties of the relation of spatial

Anthony Quinton, 'Spaces and Times', *Philosophy*, 37 (1962): 130–47. Reprinted by permission of the author and the Royal Institute of Philosophy. The text as reprinted here includes some small changes made by the author for this volume.

connection that everything is in one and the same space, that everything is spatially connected to everything else. What does follow is that if A and B are spatially connected then everything spatially connected to A is spatially connected to B and vice versa. Spatial connection is analogous in form to identity in colour, which is also symmetrical and transitive. Provided that A and B are identical in colour everything identical in colour with A is identical in colour with B. But there are, of course, many distinct colours and so many pairs of things which, while identical in colour with some things, are not identical in colour with each other. So far, then, it is an open possibility that spatial things should be arranged in spatially connected groups—just as coloured things can be arranged in colour-identical groups—all of whose members were spatially connected to each other but none of whose members were spatially connected to any of the members of any other group. To say that everything is in one space is simply to deny this possibility and to assert that all things are spatially connected. Naturally this assertion applies only to things that are in some sense spatial, things to which spatial predicates can intelligibly be ascribed. But this qualification does not affect the situation. The unity of space is not involved in the conception of a spatial thing. To say that a thing is spatial is to say either or both of the following: (*a*) that it is extended, that its parts are spatially connected to one another, and (*b*) that it is spatially related, that it is spatially connected to something distinct from itself. It does not follow from either of these or from both of them taken together that it is spatially connected to everything. It does not follow, then, from the mere conception of a spatial thing that space is a unique individual. So far the formal possibility of a plurality of spaces remains open.

The same thing holds for time, as can easily be shown. Let us call two events temporally connected if there is a time-interval between them or if they are simultaneous. This relation, like that of spatial connection, is clearly symmetrical and transitive. So it allows for self-contained and exclusive groups of temporally connected events. Nor does the unity of time follow from the conception of a temporal thing or event. A temporal thing is something that occupies a lapse of time, that has temporally connected parts or phases, and/or is something that is temporally connected to something else. From neither of these conditions does it follow that a temporal thing is temporally connected to everything.

But although the unity of space and the unity of time are not formally deducible from the concepts of spatial and temporal connection or the concepts of spatial and temporal things we do appear to believe that space and time are unities. Our direct information about the spatial and temporal connection of things is comparatively local. We observe the spatial disposition of things—the tree beside the barn, the mountain on the other side of the river—and we observe their temporal succession—the egg-white turning into a meringue, the bruise following the blow. Cartographers and chronologists piece these facts together in a single system of spatial and temporal positions. Provided that they can be answered at all, questions as to where things are or were and when they happened can always, it seems, be answered in terms of a system of positional references in which all positions are connected. As things are, if a thing cannot be found a home in this unitary system of positions we conclude that there is no such thing.

The belief in the existence of one all-embracing space and one all-embracing time has not gone unchallenged. Bradley, in his determination to show the merely apparent character of space and time, addresses himself to the question at various points in his writings.[1] He argues that the unity of space and of time is not only no necessity but that it is not even a fact. Why, he asks, should we take time as one succession and not as a multitude of series which are altogether temporally disconnected and separate although the members of each such series are temporally related to one another? In support of this proposal he draws attention to the relation between events in dreams and stories. In these imaginings events occur that are indisputably temporal entities since they are temporally related to other events in the same imagining. Yet these events cannot be located in the framework of public or historical time. Bradley rejects the suggestion that they should be dated by the time of their appearance in the mental history of the imaginer which can, we may assume, be located in ordinary public time. His argument is characteristically summary: it would be absurd, he says, to date the events of a novel by the date of its publication. The point he is making can be more persuasively developed. We can understand having good reason for saying that a dream lasted for thirty seconds or less of the dreamer's mental history while the content of the dream occupied a much greater tract of time. Here the events of the dream and the process of dreaming it are at least in the same order though the intervals between the things ordered are different. But we could also have reason for saying that the things I dreamt about on Monday were subsequent to the events I dreamed about on Tuesday. For on Monday I might have dreamt about myself as I am now and on Tuesday about myself as a child at school. Similarly it is quite possible for novelists to think up and for novel-readers to read what would naturally be called the later part of a story before the earlier part. Isherwood's *The Memorial* and Fitzgerald's *Tender Is the Night* are familiar examples of the latter possibility. Bradley goes on to suggest that even if all the events of which I am aware do fit into one all-inclusive temporal scheme it does not follow that there could not be

[1] F. H. Bradley, *Appearance and Reality* (Oxford: Clarendon Press, 1955), ch. 18, pp. 186–9.

events entirely unrelated to my time-series. But this is an empty proposal since he does not suggest any circumstances in which we could have any reason to think that there were such series. He attempts to dispose of the unity of space in a more cursory way. At first glance the order of extension seems to be one whole. But if we reflect we can see that extension is manifested in dreams: the trapeze I dream that I am swinging on is an obviously spatial thing but it is connected by no spatial route whatever to the familiar spatial contents of the common world.

Bradley's arguments for plurality all derive from the spaces and times of imagination. But since Russell's first works on the theory of knowledge we have become familiar with another source for arguments of the same kind— the spaces and times of sensation. My visual sense-data are extended, spatial entities, occupying positions and spatially interrelated to other things in the space of my momentary visual field. To the extent that my sense-data are veridical and have been obtained under normal conditions of observation they will at least correspond to the contents of common, public space. But they are not located in it. For I am the only person in the world who is even tempted to suppose that they are to be found there. And I need not give way to this temptation. If I look at a mountain and then close my eyes I do not suppose that anything at all has happened in the part of public space that is occupied by the mountain. To the extent that my sense-data are not veridical they do not usually even correspond to anything in public space. At best they have some sort of causal determinant within it. To take an example whose existence at least is uncontentious: my after-image is plainly a spatial thing, it occupies at any one moment a definite position in my visual field, but it has no real location in the public world.

There is a short but not entirely convincing answer to Bradley's arguments from the spaces and times of imagination. It could be said that imaginary objects and events, the contents of our dreams and fantasies, are nowhere at all. The contents of our imaginings are simply unreal. They can raise no problems of spatial and temporal location because they just do not exist. But to this it could be replied that although the trapeze I dreamed about last night has never hung in any actual, publicly observable circus tent, there really was something, a private entity, an image or dream-element, of which I was aware shortly before I woke up this morning. The remarks I produce at the breakfast table are not free and spontaneous creations, mere playing with words. I make earnest efforts to get my descriptions right, to leave nothing out, to set out the events dreamed of in the exact order in which I dreamt them. Bradley's argument can be countered in this way only if one is prepared to adopt a theory of dreaming like Professor Malcolm's which takes them to be no more than the utterance of sentences which, though just like the sentences we use to give genuine descriptions of our past experience, are not in fact being used for this purpose and are not intended to be understood as if they were. If this is accepted we do not have to worry about the spatial and temporal character of whatever it is that the report of a dream describes because such reports do not describe anything.

The most straightforward way of bringing out the implausibility of this theory is phenomenological: one has only to point to the experienced difference between making a story up out of one's head and reporting a dream or an earlier product of the imagination. There are two sides to the activities of the imagination: the story and the experience. The story is the words, written or spoken, in which a dream is reported or a piece of fiction is told. The experience is the body of images or private elements that the dreamer was aware of while he was dreaming, that the novelist was aware of while he was working out his book, and that the reader is no doubt intermittently aware of as he reads. Only if we can eliminate the experience, by regarding it, for example, as no more than the disposition to produce a story, can Bradley's argument be summarily disposed of. Even if we do eliminate experience from our account of imagination there is still the spatial and temporal character of sensation to be dealt with. Bradley's argument for a plurality of spaces and times can be said to rest, then, on the spatial and temporal character of private experience—of images, dreams, and sense-impressions. In private experience we are aware of things that are spatially extended and temporally enduring. These things are spatial and temporal in virtue both of the spatial and temporal relations between their parts and of the fact that they are spatially and temporally related to other spatial and temporal things. The dream-trapeze has ropes stretching away above the bar and the whole thing hovers above the sawdust surface of the dream arena. If we cannot show private experience to be the disposition to speak in a peculiar way we must either accommodate its contents in the unitary space and time of the common world or concede Bradley's point—that there is in fact a plurality of spaces and times.

We should perhaps reconsider this first alternative that Bradley so rapidly brushed aside. Can the tiger I am now picturing in my mind's eye be accommodated in public space? It cannot be accommodated at the place at which it looks as if it were. In the first place it may not look as if it were anywhere in particular. The background against which I am now experiencing it may be too dim and vague to provide any clue to location or it may be entirely unfamiliar. In these circumstances there can be no such activity as trying to find out where it is imagined to be. All I can do is to imagine it to be definitely somewhere, perhaps on the steps of the Albert Memorial and thus against a definite background located in public space. But in doing this I have not so much found out where it was as moved it there or, should one say, imagined another, no doubt very similar, tiger to be there. The situation is no better if the tiger does definitely look as if it were at some known and familiar place. For even if I dream of a tiger on the steps of the Albert Memorial, the real steps of the real Albert Memorial are not occupied by the tiger I am aware of in my dream. I can perfectly well have such a dream and accept reliable testimony that no tiger has been seen anywhere near the place I dreamt of. Even if, by some wild chance, there was an escaped tiger on the actual steps at the moment I was having my dream, we do not have to say that it is the very tiger I was dreaming of, however close the similarity. If my dreams turned out to be consistently correct representations of what was currently going on in the places I was dreaming about we might come to regard them as visions or cases of long-distance perception. But in that case they would no longer be dreams and it is characteristic of dreams that they do not exhibit any reliably attested correspondence of this kind.

The only other alternative is to locate the dreamed-of or imagined tiger at the place where I am, as, for example, quite literally, in my head. But this is an obviously hopeless manœuvre. When I dream of tigers there generally are no tigers anywhere near where I am, my head is not large enough to contain tigers, the possible pattern of electrical activity in my brain associated with dreaming of tigers is not identifiable with the tigers I dream of since I know that I have dreamt of tigers but the electrical activity is an unstable compound of hearsay and guesswork. We cannot literally identify the places *in* my experience with the places *of* my experience.

2

Is the same thing true of time? Earlier, developing a rather sketchy argument of Bradley's, I suggested that the lapse of time someone dreams of might be much greater than the interval between the time at which he began to dream of it and the time at which he stopped doing so and I also suggested that one could dream of events happening in an order opposite to that of the events of dreaming of them. It might seem that these suggestions could be resisted. Could we not say that the estimate of time made in the dream is just a mistaken one, that the dreamed-of fall from the top of the building and the dreamed-of splash into the river are really oñly a fifteenth of a second apart even though they seemed in the dream to be separated by an interval of several seconds or even minutes? Again if I dream on Monday of taking off bandages to find my wound almost healed and on Tuesday of receiving the wound with a great deal of associated connecting tissue to link the two dreams together am I not compelled to say that the time in the dreams is in reverse order to the time of dreaming them? The wound I dream of on Tuesday could be said to be a new wound in the same place, if we felt compelled to link the two dreams together. In the first example the determination to identify the time of the dream with the time of dreaming is rather gratuitous. The correlation of dreams with their manifestations in the public world is tentative and infrequent. My audible cry of 'help' does not have to be taken as simultaneous with my dreaming of a fall from the building nor my visible shudder with the splash. In most cases there is nothing even to suggest to us that the time of the events dreamed of is anything but what it appears to be. In the second example the situation is not so clear. In the first place the temporal propriety being defended is of a more fundamental kind. The topology of time order is more sacred than the geometry of time-intervals. All the same it would seem unreasonable to deny that there was a difference between the time of the dream and the time of dreaming it if on a series of twelve nights one dreamt and remembered in precise detail a series of occurrences whose content could be naturally arranged only in exactly the reverse order to that in which they were dreamed. My general conclusion, so far, then is that we do have reason for admitting the existence of a plurality of experiential spaces over and above the space of the common world and that we could have reason for a similar admission about experiential times. There is no obvious contradiction in saying that there is such a plurality and, given the implausibility of strictly verbal accounts of private experience, better reason for saying that they do exist than that they do not. However, if we consider the character of these experiential spaces and times more closely it will appear that they are so different from physical space and time that the concession we have made to Bradley's line of thought involves only a small modification of the common conviction of spatial and temporal unity.

3

There are two fundamental differences between physical and experiential space and time. Where the physical is vast and systematic, the experiential is small and fragmentary; where the physical is public, the experiential is private. These are not exactly contingent features even though comprehensiveness and publicity could vary in degree. Consider the space of dreams. There is ordinarily no ground for saying that the space of Monday night's dream has anything to do, or is any way connected, with the space of Tuesday night's dream. We often have several spatially disconnected dreams in one night. And in the course of one more or less continuous dream it is only the comparatively momentary spatial relationships of the dreamed-of things, the spatial relationships revealed in a temporal cross-section of the whole dream, that are at all definite. First I am on the trapeze. Below me I see a familiar face. Shortly afterwards my friend and I are seated side by side in a boat. Such continuity as there is is provided by the familiar face but it is not sufficient to establish any spatial relation between the trapeze and the boat. Many dreams are more coherent than this, of course, but it would seem that the constructibility of non-momentary spaces, spaces that endure as the scenes of comparatively protracted change, is the exception rather than the normal case in the experiential realm. We do have dreams where temporally successive incidents occur against a fairly definite and persisting background and there can be enough correspondence between the spatial contents of two quite distinct dreams for it to be reasonable to regard both as relating to one and the same spatial order. But as things are, this is about as much in the way of system and coherence as our dreams ordinarily yield. It is plain that the same thing holds for our imaginings, which, being so much more interrupted, so much more exposed to the solicitations of the external world, are perhaps even less coherent than dreams. It is also to some extent true of our sensations, which become coherent only as a result of a good deal of suppression and filling-in. Privacy is as obvious a feature of the experiential realm as fragmentariness or incoherence. Nobody, as things are, can tell what our dreams or imaginings are unless we tell them. In the case of sensations reliable inferences can be made on the basis of well-established correlations between sense-experience and the condition and environment of the observer. If such correlations were available for inference to the other domains of experience they would have to rest in the end on the admissions of observers. We can imagine circumstances in which the correspondence between the dreams of two or more people was so extensive as to lead us to say that they were dreaming the same dream, especially if in the event of some marginal disagreement between two corresponding dreamers one of them subsequently admitted that he was mistaken. If there were many more blind people than there are, the remarks of the sighted about clouds and sunsets might well appear to be the by-products of a widely shared dream. This is not altogether unfamiliar ground. The raptures of mysticism and musical appreciation incite, in rather different ways, just such a response amongst the less respectful of the uninitiated. Only if the correspondence becomes general enough to count as normal can dreaming come to be accounted as observation.

These differences between physical and experiential space and time are substantial enough as things are for the thesis that space and time are unitary to survive Bradley's arguments almost intact. Instead of saying that there is only one space and only one time the defender of unity must say that there is only one space and only one time that is coherent or public or both. Coherent and/or public space and time are, he might say, the only real space and time. Other spatial and temporal entities are fragmentary and private, a sort of ontological litter to be bundled into the wastepaper basket of the imaginary. He could argue that we count only those things as real that can be fitted into the one coherent and public space and time, that such locatability is a criterion of being real. For what is a dream or a fantasy or an illusion of the senses but an experience that fails to fit into the unitary spatio-temporal scheme? From this it follows that Descartes's hypothesis that perhaps everything is a dream is illegitimate. It cannot be significantly affirmed since to call a tract of experience a dream is to say that it fails to conform to the standard of coherence and publicity exhibited by the greater part of our experience. So to say that all our experience is a dream is to say that none of it comes up to the standard of most of it, a straight self-contradiction. Here is one case, at any rate, where the paradigm-case argument works. It does not entirely dispose of the Cartesian hypothesis. A man might have acquired a standard of coherence from somewhere else, perhaps a religious experience, though this would not show that the present distinction between waking and dreaming was improper, only that it should be differently named. More important is the fact that even if life hitherto cannot all have been a dream it does not follow that the whole structure will not come to pieces in the next few minutes, that from then on none of our experience will attain the standard we have come to expect and all will be as incoherent as what we have hitherto regarded as dreams.

4

The position we have arrived at, then, is that even if it is not true that absolutely everything can be located in one space and one time, everything real, provided that it is spatial and temporal at all, can be so located. If the suggestion that such locatability is a criterion of being real is correct, it follows that the thesis of unity in its revised form is a necessary truth. Now this is essentially the opinion of Kant.[2] Space and time, he said, are not discursive or general concepts of the relations of things in general but pure intuitions. In other words they are not universals but particulars and unique particulars at that. His argument is that we can conceive limited spaces and times only as parts of one all-inclusive space and one all-inclusive time. These unique particulars are not literally composed of perceptually observed

[2] *Critique of Pure Reason*, trans. N. Kemp Smith (London: Macmillan, 1929), 69, A25 = B39.

spatial extents and temporal durations, are not constructions from these extents and durations as elements, because they are somehow presupposed by
these elements.

The logical status of arguments from conceivability is always insecure and
in this type of case especially so, for we are concerned with a very primordial
feature of our experience. Our habits of thinking about space and time are so
early acquired and so deeply ingrained that their extreme familiarity can
easily look like logical indispensability. It is clear anyway that we do in fact
take all real spatial extents and temporal durations to be parts of the one space
and the one time. But Kant is claiming more than this and to assess his claim
we must ask whether we are compelled to think in this way. We can even
concede that on our present interpretation of 'real' the statement 'Everything
real is in one space and in one time' is analytic. The question still remains
whether there are any conceivable circumstances in which it would be reasonable to modify this interpretation. For it can be maintained that there are,
in a sense, degrees of analyticity. That we have a certain concept at all can
often be explained by referring to facts which might not have obtained. With
any one concept there may be a number of such explanatory facts which can
be arranged in some order of importance. The essentials of the concept would
remain if some of the less important facts did not obtain and if, therefore, the
conventions that depend on them did not exist. Let us take a very simple
example, that of brotherhood. Our existing concept of brotherhood is determined by facts of biology and sociology. Men are borne by women, as a
result of sexual intercourse between those women and other men, and pass
the helpless years of infancy in a group commonly led, protected, and provided for by their parents. Now imagine a society in which women were
elaborately promiscuous or in which all conception came about through
artificial insemination by anonymous donors. Suppose also that the family
group consisted of the mother and her children alone. In these circumstances
we should presumably count children of the same mother as brothers and the
statement 'All brothers have both parents in common' would be no longer
analytic but contingent and false. It is too narrow to describe this situation
as one in which we should use the word 'brother' to mean what we now mean
by 'maternal half-brother'. For what is really important about the concept of
brotherhood, that it relates persons who share both a biological inheritance
and certain fundamental loyalties and affections, is still retained by the
revised concept. Now suppose that children were taken from their mothers
at birth and brought up in institutions. Even here there might be some point
in having the concept if the institutions in which children were brought up
had something of the emotional structure of the ordinary human family as it
now exists.

Let us look at a more complicated and perhaps more philosophically interesting example considered by Professor Ayer, which concerns the privacy of pain.[3] As things are, the causal conditions of pain are commonly found in the body of the sufferer. If I am in pain it is not usually the case that anyone near me has a similar affliction and I cannot generally get rid of the pain by moving about. Now suppose that circumstances were different, that everyone whose body is in a certain region of space during a certain period of time feels a pain of much the same sort, that the intensity of this pain uniformly diminishes as they move away from a determinable point in the region, and that it disappears altogether when they are at a certain distance, roughly agreed upon by all, beyond this central point. In these circumstances, Ayer suggests, we might well cease to think of pains, as we now do, as being private and might come to accord them much the same sort of status as we now give to material things. 'Look out,' we might say to a man walking in a certain direction, 'there's a pain there'; and we might say this with good reason even if there were at the time no one in the region in question and therefore no one suffering the pain. If this were to come about people might cease to speak of 'my pain' and 'your pain' and there would be no question that different people could feel the very same pain. In other words the statement 'No one but me can feel the pain I am feeling' would no longer be analytic. The same thing would happen to the statement 'All pains are felt by somebody'.

It would still, of course, be open to philosophers to talk about pain-data and they might well be encouraged to do so if there were perceptible differences of sensitivity between people or if some people felt pain in places where nobody else did. They would have the same reasons for talking about pain-data and pain-hallucinations as they now have for talking about sense-data and hallucinations of the senses. Ayer's supposition reveals the contingencies on which our current convictions about the concept of pain rest. If it came true it would be reasonable to alter these conventions and to regard many statements as synthetically true or false which we now regard as analytic or contradictory. The essentials of the concept have not been tampered with; under his supposition there are still experiences which people generally and instinctively dislike having.

Can we construct a myth that will reveal the ultimately conventional character of the Kantian thesis that real space and time are unitary? Do our current convictions about the unity of space and time rest in the end on contingencies which we can conceive as ceasing to obtain? I believe that there is an important asymmetry in this respect between space and time and I shall argue

[3] A. J. Ayer, *The Problem of Knowledge* (Harmondsworth: Penguin, 1956), ch. 5, sect. 3.

that a coherent multi-spatial myth can be envisaged but not a coherent multi-temporal one. So I shall begin with space.

5

Now suppose that your dream-life underwent a remarkable change. Suppose that on going to bed at home and falling asleep you found yourself to all appearances waking up in a hut raised on poles at the edge of a lake. A dusky woman, whom you realize to be your wife, tells you to go out and catch some fish. The dream continues with the apparent length of an ordinary human day, replete with an appropriate and causally coherent variety of tropical incident. At last you climb up the rope ladder to your hut and fall asleep. At once you find yourself awaking at home, to the world of normal responsibilities and expectations. The next night life by the side of the tropical lake continues in a coherent and natural way from the point at which it left off. Your wife says 'You were very restless last night. What were you dreaming about?' and you find yourself giving her a condensed version of your English day. And so it goes on. Injuries given in England leave scars in England, insults given at the lakeside complicate lakeside personal relations. One day in England, after a heavy lunch, you fall asleep in your armchair and dream of yourself, or find yourself, waking up in the middle of the night beside the lake. Things get too much for you at the lakeside, your wife has departed with all the cooking-pots, and you suspect that she is urging the villagers to sacrifice you to the moon. So you fall on your fish-spear and from that moment on your English slumbers are disturbed no more than in the old pre-lakeside days.

There are some loose ends in this story but I think they can be tidied up. What, first of all, about your lakeside life before the dream began? Either the lakesiders will have to put up with the fact that you have lost your memory, and we can leave it open whether they are in a position to fill in the blank for you or not, or you might find 'memories' of your earlier lakeside career spontaneously cropping up. The most immediately digestible possibility is perhaps a version of the latter in which your lakeside past gradually comes back to you after an initial period of total amnesia. But complete loss of memory is the easiest to handle. Next, how are the facts that you are awake sixteen hours and asleep eight hours in each environment to be reconciled? How can sixteen hours of England be crammed into eight hours of lake and vice versa? Well, why not? As long as there is some period of sleep in each day in each place there is room for the waking day in the other place. We often say, after all, that dreams seem to take much longer than they actually do. The same principle could be applied to our alternative worlds. To make the thing fairly precise we could correlate hours in England with hours by the lakeside, on the basis of nocturnal mumblings and movements, so that midnight to 8 a.m. in England is 8 a.m. to midnight at the lake and vice versa. This would have mildly embarrassing consequences but not contradictory ones. If I stay up till 4 a.m. in England I cannot wake up beside the lake until four in the afternoon. If an alarm clock wakes me two hours early in England, i.e. at 6 a.m., then I shall find myself dropping irresistibly off at 8 p.m. by the lake. One embarrassment is common to both hypotheses: if in either place I stay up all night I must sleep all through the day in the other. Some of these embarrassments can be avoided by supposing that the lakeside day is normally eight hours long and the lakeside night sixteen hours long. To imagine such a comatose manner of life is perhaps easier than having to put up with the embarrassments of rigid correlation.

Now if this whole state of affairs came about it would not be very unreasonable to say that we lived in two worlds. So far it may seem that only one of the properties of physical space as we understand it has been added to the space of dreams, namely its coherence. But it only takes a small addition to equip it with publicity, an addition already implicit in the fable as I have told it. For I am not alone at the lakeside, there are my wife and the moon-worshipping villagers, whose statements and behaviour may confirm all the spatial beliefs I form at the lakeside, with the usual minor exceptions. It might be argued that this sort of publicity is bogus, that it is only dream-publicity. But as it stands this is just prejudice. At the lakeside, on my hypothesis, we have just as good reason to take our spatial beliefs as publicly confirmed as we have in England. However, a less questionable type of publicity can be provided if we suppose that the dreams of everyone in England reveal a coherent order of events in our mythical lake district and let everyone have one and only one correlated lake-dweller whose waking experiences are his dreams. (In this case we should have to correlate the clocks of England and the lakeside, either by the rather embarrassing proposal of elastic time-intervals or by that of the eight-hour lakeside day. For otherwise I could drop off at the lakeside on Monday, wake up before you go to sleep, and tell you a whole lot of things about Tuesday in England before, from your point of view, they had happened.) There are various ways in which we can suppose that people who know one another in England could come to recognize one another at the lakeside, for example, by the drawing of self-portraits from memory or by agreeing, in England, to meet at some lakeside landmark.

I shall not pursue the hypothesis of a dream that is public in this strong sense since it becomes imaginatively too cumbrous, though not, I think, self-contradictory. One special difficulty is that in such circumstances some measure of causal interpenetration by the two worlds would be natural. Even if physical causation cannot, *ex hypothesi*, operate from one region into the other, psychological causation presumably would. The injury I do you at the lakeside may be revenged not there but in England. I think, in fact, that my original one-man hypothesis is sufficient for the purpose in hand. But we can publicize this a bit further without going all the way to publicity in the strong sense. It might, for instance, be the case that everyone's dream-life was coherent but that no one person's dream-life corresponded with anyone else's. In this case everyone would inhabit two real spaces, one common to all and one peculiar to each. This residual asymmetry can, of course, be eliminated by requiring that the same be true of all the lakesiders. On this supposition the worlds we have some reason to believe in fray off into infinity. Each of my lakeside acquaints has his other life, in which he comes across people, each of whom has his, and so on. But this infinity does not seem to be a vicious one.

It might be said that if this myth were realized we should either have to say that the dream-place was somewhere in ordinary physical space or else that it was still only a dream. Both of these alternatives can be effectively disputed. Suppose that I am in a position to institute the most thorough geographical investigations and however protractedly and carefully these are pursued they fail to reveal anywhere on earth like my lake. But could we not then say that it must be on some other planet? We could but it would be gratuitous to do so. There could well be no positive reason whatever, beyond our fondness for the Kantian thesis, for saying that the lake is located somewhere in ordinary physical space and there are, in the circumstances envisaged, good reasons for denying its location there. Still suppose we do find a place, in New Guinea let us say, exactly like the lakeside. I, the dreamer, lead the expedition into the village, brandishing trade goods. Friendly relations having been established with the elders of the place, we are led to the longhouse to meet the populace and there, to my amazement, is the face I have often seen in my dreams while bending over the pools at the lake's edge in pursuit of fish. Now if the owner of this face is fast asleep and cannot be woken up until I go to sleep this village and the place of my dreams can be identified. But suppose he is wide awake and we get into conversation. He turns out to have coherent dreams about my life in England and in fact to have dreamt last night of my progress towards the village on the preceding day, just as I dreamt last night of what he was doing the day before. The natural conclusion will be that we are connected by a kind of delayed cross-telepathy and that what the Kantian insists are still only dreams are at any rate in the same order of reality as dreams. If, then, we do find what is to all intents and purposes the place of my dreams, the Kantian's dilemma—either

in the one real space or just a dream—does apply. But if we do not there is
no reason to insist upon it.

If, failing to find the scene of my coherent dream in ordinary physical
space, we insist that it is, then, only a dream we are neglecting the point of
marking off the real from the imaginary. Why, as things are, do we have this
ontological wastepaper basket for the imaginary? Because, approximately,
there are some experiences that we do not have to bother about afterwards,
that we do not, looking back on them, need to take seriously. Dream-events,
where they have consequences at all, do not have serious consequences. If I
dream of cutting somebody's throat my subsequent dreams will in all probability be entirely unrelated to him and to my act. Even if they are, when I
am haled into court I am as likely to be given a bunch of flowers as a death-
sentence. But beside the lake there is a place for prudence, forethought, and
accurate recollection. It is an order of events in which I am a genuine agent.
There is every reason there for me to take careful note and make deliber-
ate use of my experience. Reality, I am suggesting, then, is that part of our
total experience which it is possible and prudent to take seriously. It is, of
course, because I am ultimately interpreting reality in this way that I can
envisage dispensing with locatability in one physical space and time as a
criterion of it. My conclusion so far, then, is that it is a contingent matter that
the experience we can and prudently should take seriously can all be assigned to one space. Kant's unity of space is not an unalterable necessity of
thought.

6

Let us turn finally to the case of time. Can an analogous myth be constructed
here? Can we conceive of living in two distinct orders of spatial extension?
The lakeside story did present some peculiar temporal features, in some of
its versions a sort of time-stretching, but at least the proprieties of temporal
order were respected. And, with the eight-hour day, it was possible to do
without time-stretching. Another avoidable difficulty was the temporal status
of the events 'remembered' by me at the lakeside after I have started, *nel
mezzo del cammin*, consciously living there. Here again temporal order is all
right. The trouble arises about the correlation of my 'remembered' twelfth
birthday and initiation ceremony at the lakeside with events in my English
life. However, this is not a serious problem. Either we can say the date of
my initiation in English terms is unknown, apart from being before such and
such a date on which my lakeside experiences started or we can extrapolate
with the help of rules of correlation we have established in the directly

experienced parts of my lakeside life. Our multi-spatial myths are not, then, also multi-temporal myths.

So for a multi-temporal myth we must begin again from the beginning. What we are in search of, in general terms, is this: two groups of orderly and coherent experiences where the members of each group are temporally connected but no member of either group has any temporal relation to any member of the other. Such a search seems doomed from the start. How can these experiences be my experiences unless they constitute a single temporal series? This will become clearer if we consider some examples of possible multi-temporal myths.

Consider first the myth that results from a small complication of the original myth about England and the lakeside. Suppose that my memories become, so to speak, disconnected, that I can remember the relative temporal situation of English events and the relative temporal situation of lakeside events but not the temporal relations of any English events to any lakeside event. I can remember that I got on the bus after I had spoken to Jones about our favourite television programme. I can remember taking part in the fertility-rite after setting the fish-traps. What I cannot remember is whether getting on the bus occurred before or after setting the fish-traps. The trouble with this obstacle to unitary dating is that it is too easily circumvented. At the beginning of day 1 in England I write down in order all the lakeside events I can remember. On day 2 in England I cannot remember whether the events of day 1 followed or preceded the lakeside events in the list. But the list will be there to settle the matter and I can, of course, remember when I compiled it.

A desperate shift that might suggest itself at this point is the supposition that I cannot remember lakeside events at all when I am in England nor English events while I am at the lakeside. But this is self-destroying. For unless I have memories of one series of events while experiencing the other there can be no reason for saying that I am involved in both of them, that both are experienced by one and the same person. *Ex hypothesi* the lakeside can have no physical, observable traces in England, so my memories of it in England are the only reason there can be for me, in England, to think that the lakeside exists.

Another line of approach requires us to suppose that the experience of dreaming coherently about the lakeside is general or at least widespread. It might be thought that we could all pass in and out of the coherent dream-world, or the alternative reality, at different times. But suppose the salient events of the day for two people in England are kipper for breakfast and steak for lunch, while the salient event for the approximately corresponding day at the lakeside is a distant volcanic eruption. The two people breakfast together in England, so their kipper-eating is simultaneous. After breakfast one of them drops off and witnesses the volcanic eruption. He awakes for lunch and over their simultaneous steaks tells his partner about the eruption. After lunch the second man falls asleep and witnesses the eruption for himself. At first glance this might seem to suggest that the eruption cannot be fitted at all into the English time-sequence. But on reflection it is clear that we can fit 'the eruption' in only too well. For it happened, to A, before their simultaneous lunch and, to B, after it. What happens before an event, happens before everything that happens after that event. Therefore, the eruption happened before itself. The only consistent conclusion from the data is that two eruptions took place and for each of these there is a perfectly unequivocal position in the English time-series.

The moral of these unsuccessful attempts to construct a multi-temporal myth is the same in each case. Any event that is memorable by me can be fitted in to the single time-sequence of my experience. Any event that is not memorable by me is not an experience of mine. This second proposition is not equivalent to a Lockean account of personal identity which holds my experiences to be all those experiences that I can remember. For the memorability to which it refers is memorability in principle not in practice. All that is required for an experience to be mine is that I should be logically capable of remembering it. But from the fact that, at a given time, I am logically capable of remembering a certain experience, it follows that the experience is temporally antecedent to the given time, the time of my current experience, and so is in the same time, the same framework of temporal relations, as it is. Thus if an experience is mine it is memorable and if it is memorable it is temporally connected to my present state. The question we are raising—is it conceivable that we should inhabit more than one time—answers itself. For what it asks is: could my experience be of such a kind that the events in it could not be arranged in a single temporal sequence? And it seems unintelligible to speak of a collection of events as constituting the experience of one person unless its members form a single temporal sequence. This view of the concept of a person's experience is supported by another consideration. It is possible to imagine that our experience might not be spatial. As Mr Strawson has shown, if our experience were all auditory, although it might contain features and differentiations which could be used as clues to spatial position with the aid of correlations with the deliverances of other senses, these features would have no spatial import on their own.[4] On the other hand it is not possible to imagine an experience that is not temporal. We should, of course, have no sense of the passage of time unless

---

[4] P. F. Strawson, *Individuals* (London: Methuen, 1959), ch. 2.

our experience exhibited change. But an unchanging experience is no more intelligible than a non-temporal one. An experience of one unvarying sound, or even of an unvarying mixture of sounds, would not be an experience at all. A high, thin, metaphysical whistle sounding in one's mind's ear from birth to death would be in principle undetectable, like the impression of the self that Hume rummaged unsuccessfully around in his consciousness for.

I conclude, then, that we can at least conceive circumstances in which we should have good reason to say that we knew of real things located in two quite distinct spaces. But we cannot conceive of such a state of affairs in the case of time. Our conception of experience is essentially temporal in a way in which it is not essentially spatial.

NOTES ON THE CONTRIBUTORS

MICHAEL DUMMETT is Wykeham Professor of Logic at the University of Oxford and a Fellow of New College; between 1968 and 1984 he was a Fellow of the British Academy. His publications include *Frege: Philosophy of Language* (1973; 2nd edition, 1980), *Elements of Intuitionism* (1976), *The Interpretation of Frege's Philosophy* (1981), *Voting Procedures* (1984), and *Frege and Other Philosophers* (1991). A collection of his essays, *Truth and Other Enigmas*, appeared in 1978.

GRAEME FORBES is Professor of Philosophy at Tulane University, New Orleans. He is the author of *The Metaphysics of Modality* (1985), *Languages of Possibility* (1989), and the forthcoming textbook *Modern Logic*.

ROBIN LE POIDEVIN is Lecturer in Philosophy at the University of Leeds, and was earlier Gifford Research Fellow at the University of St Andrews. He is the author of *Change, Cause and Contradiction: A Defence of the Tenseless Theory of Time* (1991).

DAVID LEWIS is Professor of Philosophy at the University of Princeton. His publications include *Convention* (1969), *Counterfactuals* (1973), *On the Plurality of Worlds* (1986), and *Parts of Classes* (1991). Two volumes of his collected essays have appeared: *Philosophical Papers*, volume i (1983) and volume ii (1986).

MURRAY MACBEATH is Lecturer in Philosophy at the University of Stirling. He has published many articles on the philosophy of time, moral philosophy, and philosophical theology.

J. M. E. MCTAGGART, who died in 1925, was Fellow of Trinity College, Cambridge, from 1891 to 1922. His best-known work, from which his essay in this volume is drawn, is *The Nature of Existence* (volume i, 1925; volume ii, 1927).

D. H. MELLOR is Professor of Philosophy at the University of Cambridge, a Fellow of Darwin College, and a Fellow of the British Academy. He is the author of *The Matter of Chance* (1974) and *Real Time* (1981). A volume of some of his essays, *Matters of Metaphysics*, was published in 1991.

W. H. NEWTON-SMITH is Fellow of Balliol College, Oxford, and Praefectus of Holywell Manor. He is the author of *The Structure of Time* (1980), *The Rationality of Science* (1981), and *Logic: An Introductory Course* (1985).

ARTHUR N. PRIOR, who died in 1969, was Fellow of Balliol College, Oxford, and before that Professor of Philosophy at the University of Manchester. He wrote extensively on logic and the philosophy of time, and his publications included *Time and Modality* (1957), *Past, Present and Future* (1967), and *Papers on Time and Tense* (1968), with more of his work having been published posthumously.

ANTHONY QUINTON, now Lord Quinton, FBA, was Fellow of New College, Oxford at the time of writing his contribution to this volume. He went on to become President of Trinity College, Oxford, Chairman of the Board of the British Library, and Vice-President of the British Academy. His publications include *The Nature of Things* (1973), *Utilitarian Ethics* (1973), and *Thoughts and Thinkers* (1987).

SYDNEY SHOEMAKER is Susan Linn Sage Professor of Philosophy at Cornell University. His publications include *Self-Knowledge and Self-Identity* (1963) and *Identity, Cause, and Mind* (1984), which is a collection of his own essays. He has also co-edited *Knowledge and Mind* (1982) with Carl Ginet and co-authored *Personal Identity* (1984) with Richard Swinburne.

LAWRENCE SKLAR is Professor of Philosophy at the University of Michigan. He is the author of *Space, Time and Spacetime* (1974) and *Philosophy and Spacetime Physics* (1985).

ANNOTATED BIBLIOGRAPHY

D. H. Mellor's *Real Time* contains a fairly comprehensive bibliography of writings, both books and articles, on the philosophy of time published between about 1900 and 1980. With only one exception the date of first publication of the items in our bibliography is 1970 or later, but the emphasis is on work published since 1980.

Complete details of books mentioned here are given at the end of the Bibliography.

PART 1: TIME AND TENSE

The view of McTaggart himself (in Essay I) and of Mellor (in Essay III) that McTaggart's infinite regress is vicious is criticized by Quentin Smith in 'The Infinite Regress of Temporal Attributions' (*Southern Journal of Philosophy*, 24 (1986): 383–96). For a debate on the interpretation of McTaggart, see Kenneth Rankin's 'McTaggart's Paradox: Two Parodies' (*Philosophy*, 56 (1981): 333–48) and George Schlesinger's 'Reconstructing McTaggart's Argument' (*Philosophy*, 58 (1983): 541–3). Schlesinger's *Aspects of Time* also has useful material on McTaggart, and a lively discussion of the extent of the similarities between time and space.

A. N. Prior's 'The Notion of the Present' (*Studium Generale*, 23 (1970): 245–8) defends a view which develops out of ideas in his Essay II in this volume and which can be called 'temporal solipsism' ('the present', Prior says, 'simply *is* the real').

The question of the status of tense has been central in most recent work on the philosophy of time. Influenced by considerations best set out in John Perry's 'The Problem of the Essential Indexical' (*Noûs*, 13 (1979): 3–21), tenseless theorists have developed what is known as 'the new tenseless theory of time', according to which tensed statements, though they cannot be translated into tenseless ones, have truth-conditions statable in a tenseless meta-language: in short, the *A* series is ontologically, though not linguistically, reducible to the *B* series. D. H. Mellor (see his book *Real Time*, from which Essay III is drawn) is the leading exponent of this kind of theory; two others, who relate it specifically to the question of the passage of time, are J. J. C. Smart ('Time and Becoming', in Peter van Inwagen (ed.), *Time and Cause*, 3–15) and Michelle Beer ('Temporal Indexicals and the Passage of

Time', *Philosophical Quarterly*, 38 (1988): 158–64). The theory is criticized by Quentin Smith in 'Problems with the New Tenseless Theory of Time' (*Philosophical Studies*, 52 (1987): 371–92) and defended against Smith's objections by L. Nathan Oaklander in 'A Defence of the New Tenseless Theory of Time' (*Philosophical Quarterly*, 41 (1991): 26–38). Further recent criticism of the new tenseless theory is in Richard Swinburne's 'Tensed Facts' (*American Philosophical Quarterly*, 27 (1990): 117–30). See also Graham Priest's 'Tense and Truth Conditions' (*Analysis*, 46 (1986): 162–6) and Mellor's reply, 'Tense's Tenseless Truth Conditions' (*Analysis*, 46 (1986): 167–72).

A tenseless theory of time is defended in Robin Le Poidevin's *Change, Cause and Contradiction*, and one classic problem for such theories, Prior's 'thank goodness that's over' problem, is addressed in Murray MacBeath's 'Mellor's Emeritus Headache' (*Ratio*, 25 (1983): 81–8), to which Mellor replied with 'MacBeath's Soluble Aspirin' (*Ratio*, 25 (1983): 89–92).

On the passage of time, see further Ferrel Christensen's 'The Source of the River of Time' (*Ratio*, 18 (1976): 131–43), George Schlesinger's 'How Time Flies' (*Mind*, 91 (1982): 501–23), Murray MacBeath's 'Clipping Time's Wings' (*Mind*, 95 (1986): 233–7), and Arnold B. Levison's 'Events and Time's Flow' (*Mind*, 96 (1987): 341–53). On the status of the present see Jeremy Butterfield's 'Seeing the Present' (*Mind*, 93 (1984): 167–76), and on that of the future see Robert Merrihew Adams's 'Time and Thisness' (*Midwest Studies in Philosophy*, 11 (1986): 315–29). On the relevance of special relativity to the status of the present see D. H. Mellor's 'Special Relativity and Present Truth' (*Analysis*, 34 (1974): 74–8), William Godfrey-Smith's 'Special Relativity and the Present' (*Philosophical Studies*, 36 (1979): 233–44), and Lawrence Sklar's 'Time, Reality and Relativity' (in Richard Healey (ed.), *Reduction, Time and Reality*, 129–42).

PART 2: RELATIONISM ABOUT TIME

The single most sustained treatment of the debate between absolutism and relationism is John Earman's *World Enough and Space-Time*. This is a detailed discussion of the history of the debate, as well as a contribution to contemporary debate about the foundations of the general theory of relativity. Indeed, much of the most recent discussion of relationism presupposes familiarity with relativity theory. J. L. Mackie provocatively argues for absolutism and against the view that special relativity supports relationism in 'Three Steps towards Absolutism' (in Richard Swinburne (ed.), *Space, Time and Causality*, 3–22). Clifford A. Hooker's 'The Relational Doctrines of Space and Time' (*British Journal for the Philosophy of Science*, 22 (1971): 97–130) is a very thorough examination of the implications of relational theories of space, of time, and of space-time. A much more accessible, and briefer, treatment is W. H. Newton-Smith's 'Space, Time and Space-Time: A Philosopher's View' (in Raymond Flood and Michael Lockwood (eds.), *The Nature of Time*, 22–35). On modal relationism, see chapter 3 of Sklar, *Space, Time, and Spacetime*, Jeremy Butterfield's 'Relationism and Possible Worlds' (*British Journal for the Philosophy of Science*, 35 (1984): 101–13), and Paul Teller's 'Substance, Relations and Space-Time' (*Philosophical Review*, 100 (1991): 363–97).

On the possibility of time without change, see section 2 of J. R. Lucas's *A Treatise on Time and Space*, chapter 2 of W. H. Newton-Smith's *The Structure of Time*, which extends the argument of Shoemaker in Essay IV, S. G. Williams's 'On the Logical Possibility of Time without Change' (*Analysis*, 46 (1986): 125–8), and chapter 6 of Le Poidevin's *Change, Cause and Contradiction*.

PART 3: THE DIRECTION OF TIME

Paul Horwich's *Asymmetries in Time* offers clear and helpful discussion of many of the issues in the philosophy of science that bear on the problem of the direction of time.

There are useful chapters on the direction of time in W. H. Newton-Smith's *The Structure of Time* and D. H. Mellor's *Real Time*, the former criticizing, the latter advocating a causal theory. On causal theories of the direction of time, see John Earman's 'Causation: A Matter of Life and Death' (*Journal of Philosophy*, 73 (1976): 5–25), David H. Sanford's 'The Direction of Causation and the Direction of Conditionship' (*Journal of Philosophy*, 73 (1976): 193–207), David Lewis's 'Counterfactual Dependence and Time's Arrow' (*Noûs*, 13 (1979): 455–76; reprinted with a postscript in his *Philosophical Papers*, vol. ii (New York: Oxford University Press, 1986), 32–66), the symposium between Richard A. Healey and W. H. Newton-Smith, 'Temporal and Causal Asymmetry' (in Richard Swinburne (ed.), *Space, Time and Causality*, 79–121), and chapters 7 and 8 of Le Poidevin's *Change, Cause and Contradiction*.

John Earman's 'An Attempt to Add a Little Direction to "The Problem of the Direction of Time"' (*Philosophy of Science*, 41 (1974): 15–47) is a detailed examination of several distinct issues that have been taken to be relevant to the problem of the direction of time. Michael Dummett redeploys arguments about backwards causation from Essay VII, and applies them to time travel and Newcomb's paradox in 'Causal Loops' (in Raymond Flood and Michael Lockwood (eds.), *The Nature of Time*, 135–69). D. H. Mellor seeks to turn Dummett's argument in Essay VII against himself, to disprove the possibility of backwards causation, in chapter 10 of his book *Real Time* and in 'Fixed Past, Unfixed Future' (in Barry Taylor (ed.), *Michael Dummett: Contributions to Philosophy* (Dordrecht: Nijhoff, 1987), 168–88); Dummett's reply to the latter is in the same collection, pp. 287–98. Mellor's most recent statement of his causal theory of the direction of time and of his argument against causal loops is in 'Causation and the Direction of Time' (*Erkenntnis*, 35 (1991): 191–203). For a criticism of his argument against backwards causation see Peter J. Riggs's 'A Critique of Mellor's Argument against "Backwards" Causation' (*British Journal for the Philosophy of Science*, 42 (1991): 75–86).

For more on time travel, see Jonathan Harrison's 'Dr Who and the Philosophers' (*Proceedings of the Aristotelian Society*, suppl. vol. 45 (1971): 1–24), Murray MacBeath's 'Who Was Dr Who's Father?' (*Synthese*, 51 (1982): 397–430), chapter 7 of Paul Horwich's *Asymmetries in Time*, William Lane Craig's 'Tachyons, Time Travel, and Divine Omniscience' (*Journal of Philosophy*, 85 (1988): 135–50), and chapter 8 of Christopher Ray's *Time, Space and Philosophy*.

PART 4: THE TOPOLOGY OF TIME

W. H. Newton-Smith's *The Structure of Time* provides a comprehensive and accessible treatment of topological issues, covering linearity, unity, boundedness, and continuity. Newton-Smith argues in favour of regarding topological theories as empirical, because they are only contingently true or false. He takes certain fantasy arguments to establish the logical possibility of 'non-standard' topologies. For a criticism of his use of fantasy arguments, see D. H. Mellor's review, 'Theoretically Structured Time' (*Philosophical Books*, 23 (1982): 65–9), and for a comprehensive attack on the use of fantasy arguments in philosophy, see J. L. H. Thomas's 'Against the Fantasts' (*Philosophy*, 66 (1991): 349–67).

In contrast to Newton-Smith's view is that put forward in J. R. Lucas's *The Future*. Lucas attempts to derive certain topological consequences (namely, that time is linear, continuous, and non-branching) from the view that the future is unreal. For a commentary on his derivations, see Robin Le Poidevin's critical notice of *The Future* (*Philosophical Quarterly*, 41 (1991): 333–9). Richard Swinburne's *Space and Time* too has an approach to the topology of time radically at odds with Newton-Smith's; chapter 2 has a discussion of Quinton's Essay XII and the possibility of spatially unrelated spaces, while chapter 10 considers the possibility of there being two temporally unrelated time systems.

A rather technical discussion of topology is provided by John Earman's 'How to Talk about the Topology of Time' (Noûs, 11 (1977): 211–26), while Bas C. van Fraassen's *An Introduction to the Philosophy of Time and Space* has very accessible material on topology, as do the opening sections of J. R. Lucas's elegant but eccentric book, *A Treatise on Time and Space*.

Kant's famous argument for the impossibility of non-beginning time in the Antinomies is resurrected by G. J. Whitrow's 'On the Impossibility of an Infinite Past' (*British Journal for the Philosophy of Science*, 29 (1978): 39–45). For a reply see William Lane Craig's 'Whitrow and Popper on the Impossibility of an Infinite Past' (*British Journal for the Philosophy of Science*, 30 (1979): 165–70). Orientability as a property of time is discussed in John Earman's 'Kant, Incongruous Counterparts, and the Nature of Space and Space-Time' (*Ratio*, 13 (1971): 1–18). Lawrence Sklar's *Space, Time, and Spacetime* contains material on linearity, continuity, and dimensionality. There is a discussion of ancient treatments of closed time in Richard Sorabji's 'Closed Space and Closed Time' (*Oxford Studies in Ancient Philosophy*, 4 (1986): 215–31). Richard Sorabji's *Time, Creation and the Continuum* is a magisterial study of the treatment of many of the themes in this anthology, including the passage, the beginning, and the continuity of time, from ancient Greece to the early Middle Ages; as well as being a scholarly historical work, it has a keen eye for the contemporary relevance of the theories considered.

Mellor takes his argument against backwards causation (see Part 3 above) to be a decisive objection to the hypothesis of closed time. This is criticized in Susan Weir's 'Closed Time and Causal Loops: A Defence against Mellor' (*Analysis*, 48 (1988): 203–9).

BOOKS

Collections of Original Articles

Some of the articles in these collections are referred to individually above, but each of the collections also contains further relevant material.

Flood, Raymond, and Lockwood, Michael (eds.), *The Nature of Time* (Oxford: Blackwell, 1986).

Healey, Richard (ed.): *Reduction, Time and Reality* (Cambridge: Cambridge University Press, 1981).

Swinburne, Richard (ed.), *Space, Time and Causality* (Dordrecht: Reidel, 1983).

van Inwagen, Peter (ed.), *Time and Cause* (Dordrecht: Reidel, 1980).

Monographs

Earman, John, *World Enough and Space-Time* (Cambridge, Mass.: MIT Press, 1989).

Horwich, Paul, *Asymmetries in Time* (Cambridge, Mass.: MIT Press, 1987).

Le Poidevin, Robin, *Change, Cause and Contradiction* (Basingstoke: Macmillan, 1991).

Lucas, J. R., *The Future* (Oxford: Blackwell, 1989).

—— *A Treatise on Time and Space* (London: Methuen, 1973).

Mellor, D. H., *Real Time* (Cambridge: Cambridge University Press, 1981).

Newton-Smith, W. H., *The Structure of Time* (London: Routledge & Kegan Paul, 1980).

Ray, Christopher, *Time, Space and Philosophy* (London: Routledge, 1991).

Schlesinger, George N., *Aspects of Time* (Indianapolis: Hackett, 1980).

Sklar, Lawrence, *Space, Time, and Spacetime* (Berkeley, Calif.: University of California Press, 1974).

Sorabji, Richard, *Time, Creation and the Continuum* (London: Duckworth, 1983).

Swinburne, Richard, *Space and Time* (London: Macmillan, 1968; 2nd edn. 1981).

van Fraassen, Bas C., *An Introduction to the Philosophy of Time and Space* (New York: Random House, 1970).

INDEX OF NAMES